Anigrafs

Anigrafs
Experiments in Cooperative Cognitive Architecture

Whitman Richards

The MIT Press
Cambridge, Massachusetts
London, England

MIT Press books may be purchased at special quantity discounts for business or sales promotional use. For information, please email special_sales@ mitpress.mit.edu

This book was set in Sabon by Toppan Best-set Premedia Limited. Printed and bound in the United States of America.

Library of Congress Cataloging-in-Publication Data

Richards, Whitman.
  Anigrafs : experiments in cooperative cognitive architecture / Whitman Richards.
    pages cm
  Includes bibliographical references and index.
  ISBN 978-0-262-52778-1 (pbk. : alk. paper)   1. Cognition.   2. Group decision making.   3. Artificial intelligence.   I. Title.
  BF311.R487   2015
  153–dc23
                                                              2014046107

10 9 8 7 6 5 4 3 2 1

To Waltraud . . .
. . . and mentors past and present,
with special thanks to Hans-Lukas Teuber for insights leading
to new avenues for study.

Contents

Preface

Unlike the engineering systems in vogue today (e.g., ACT-R, Soar), "A Collaborative Cognitive Architecture" is a theoretical proposal, supported by simulations. The core ideas came from a fortunate convergence of three insights. The first was the realization that graph theory provided a powerful tool for understanding similarity relations among choices. The second was a new book by Donald Saari, *The Basic Geometry of Voting*. The final linchpin was work underway by Diana Richards on chaotic behavior in collective choice settings. These insights came together while I was teaching a course on cognitive architectures in 2000. The class then took on a new flavor, with emphasis on a framework incorporating decision-making among alternative choices displayed by a similarity graph. Although the approach ranged from simple brains to complex societies, the principal setting used synthetic animals with graph models for their cognitive structures.

In addition to having a formal framework for cognitive structures, graph theory allowed us to create partial orders for preferences among the alternatives. I had used this approach earlier with Alan Jepson to decide which one of all possible interpretations should be chosen for a percept. In the domain of cognitive architectures, however, the evaluation process could introduce instabilities. In 2002, Brendan McKay, Diana

Slice through the space of winners for a ring anigraf of ten vertices with two diameters at positions 8 and 10. All edges are directed. Textured regions show chaotic behavior (i.e., top-cycles) for some weights on nodes. In this case, the knowledge depth is 3. (See Anigraf2 for an explanation of knowledge depth.) Weights are (x = 0–10, 0, y = 0–10, 3, 8, 5, 1, 6, 3). For a knowledge depth of 2, position 6 in the upper-left corner occupies three-quarters of the square. For a knowledge depth of 4, three-quarters of the winners are at either position 1 or 6. Maximum complexity is at a knowledge depth of 3, above. See Appendix: Phase Plots, for construction details.

Richards, and I completed two important papers that addressed this issue and presented some conditions among relationships needed for stability when engaged in decision-making using Condorcet aggregation. (This method had been shown by H. Peyton Young to be maximum-likelihood optimal for social orders.) Our approach thus moved sharply away from Selfridge's pandemonium, "winner-take-all" method used by the vast majority in my community. This book takes the reader down the profitable highways we explored.

I would like to give many thanks to the students in my classes who provided useful feedback that helped me improve the presentation and results. Especially useful was code written by T. J. Purtell, which generated quilt-like patterns showing phase transitions between outcomes. Also greatly appreciated was the introduction by Sylvia Richards-Gerngross to the cartoonist Steven Krall. She and Nora Richards Bender lent helpful hands during the last phases of the writing of this book. Support for these studies was received to a large part from the Air Force Office of Scientific Research. I'm also indebted to Marie Lufkin Lee, Computer Science and Linguistics Editor at the MIT Press, for her practical and insightful suggestions.

W. Richards, August 20, 2014

Part I

Preliminaries: From Babble to Barter and Beyond

Decision-making is a reflection of intelligence. The lowest forms of animals make reflexive actions, whereas advanced mammals, living in complex environments, make decisions for actions that are often also complex. In such environments, many possible alternatives need to be evaluated before a commitment is made. There are two aspects of this part of intelligence that interest us here: first, there is the procedure for making a decision among many alternatives; second, there are the possible relations among these alternatives, which we dub the *cognitive architectures* of the system. These alternatives are possible choices for actions. Cognitive architectures thus capture the models the creature has built for its world, and the relations among these models. In this treatment, we cast the array of choices and models as a graph.

If we now look backward down the evolutionary tree, we see that not only are the cognitive architectures simpler (e.g., having fewer choices), but so are the choice evaluation procedures. In the lowest animal forms, a simple "loudest voice" may predominate, whereas in higher animals, the procedure may involve evaluating all relevant alternatives in a pairwise manner (known as Condorcet methods). The latter method may lead to instabilities or uncertainties in the final choice, or simply signal that no clear choice is available. These problems may be avoided in one of the least complex evaluation methods (e.g., pandemonium), where there is a single winner, which may not be optimal. Any deep understanding of intelligence needs to consider such aspects of decision-making. In these preliminaries, we provide the groundwork needed to begin such an understanding.

1.0 From Vehicles to Anigrafs

More than twenty years ago, Valentino Braitenberg wrote a fascinating and influential monograph entitled *Vehicles: Experiments in Synthetic Psychology*. The vehicle was a little robot with two rear wheels individually powered by the activity of two sensors located at the front of the robot. With very simple neural circuits, the robot exhibited movements that were naturally labeled as "Flee," "Explore," "Love," "Attack," etc., depending on the circuit design. Figure 0.1 illustrates a form of Braitenberg's vehicle, which here is transporting five daemons, each labeled according to one of the circuit designs. A daemon thus represents not only a behavior, or a cognitive objective, but also, implicitly, an underlying circuit that enables this behavior.

The mental organisms we have called daemons form a social structure, which is represented by a graph showing the similarity relations between the desires of each daemon. This graphical depiction is called an anigraf. Anigrafs have several such daemons, or mental organisms, in residence, each with its own machinery for eliciting a behavioral action.

A simplistic view of mind would be to regard a collection of vehicle-like organisms as a complex, multi-input controller. Each organism would be designed to control one variable, such as food cravings, sexual activity, foraging, and, for members of bartering societies, costs, benefits, and liabilities. The

Figure 0.1

Five daemons attempting to control the action of a bus-like vehicle.

encapsulated set of such organisms would then evaluate trade-offs over many control variables. This is the common view of how brains are organized. However, how can the craving for food be compared with sexual activity or the desire to discuss a problem with a colleague? How should degrees of risk be mapped into a pleasurable experience? Without some measure common to all choice dimensions, the typical feedback controller becomes inadequate. This leads us to consider the collection of encapsulated organisms as social decision-makers, guiding complex behaviors by reaching a collective choice. The preliminaries that follow argue for special procedures for aggregating opinions, desires, and needs of a group of such mental organisms that occupy the same physical entity.

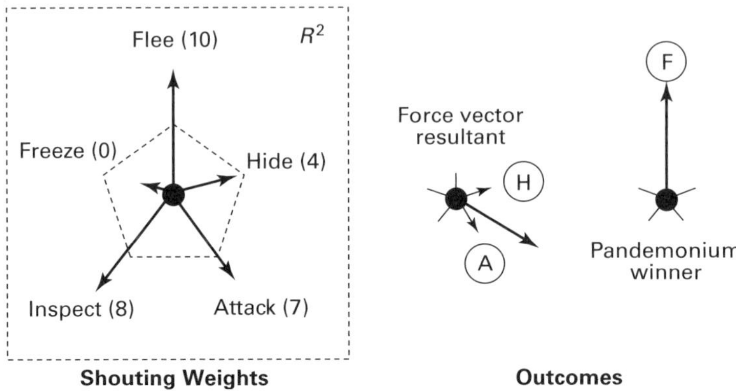

Figure 0.2

The left panel illustrates five direction vectors. The right panel shows the resultant.

1.1   Pandemonium: Babble

In vehicles there is only one daemon or mental organism dictating the action. For anigrafs there are always several. Which of the daemons should dominate the action of the anigraf? In 1958, Selfridge proposed that the one with the loudest voice would dictate the choice, the level of the shout reflecting the strength of that daemon's desire. Figure 0.2 shows possible strength values for five daemons. In this case, "Flee" is the vehicle's next move.

Another scheme, more typical of physical systems, would be to assign force vectors to each possible action, with the length of the vectors corresponding to each daemon's desire. There are two obvious advantages to this scheme. First, the resultant is easily calculated. (Think of weights at the end of each vector causing the pentagon arrangement to tilt; the location of the greatest tilt is the resultant.) Second, all daemons have a say in the final choice. In this case, when vectors are added, the resultant most often will point between categories, as illustrated in figure 0.2 (right). But half-hearted choices that meld two or more categories are inconsistent with animal behaviors. Hence our daemons will always make categorical choices, picking the direction nearest to the resultant. For our toy example, then, the group's choice would be to Attack (A), rather than to Flee (F), as depicted in figure 0.2 (left). Note also that we have our first vehicle-anigraf: the relations between the direction vectors form a simple graph, namely a pentagon.

1.2 Social Contracts

The pandemonium and vector addition methods for choosing winners are among the most simple and primitive methods to implement. They are most likely used by very primitive creatures in the natural world. More sophisticated methods for aggregating opinions require a social contract among the participants. A simple example is used in Congress: the winner is the choice that more than half of the members support. This is a plurality procedure, where the definition of a majority is 50 percent. However by agreement, a plurality could also be 66 percent or 75 percent. (In figure 0.2, note that our pandemonium example would be a plurality winner of 33 percent.) We may wish to further amend either of the above two methods to ensure that only very robust decisions are made. For example, if the plurality shout was less than one-half or two-thirds of the total volume of shouting, then no winner would be chosen, and the vehicle would simply continue with its current action. Alternately, if the resultant of the vector is either too small or lies within an angle α of the bisector of two action vectors, again no decision would be made. Adding these conditions avoids cases where the loudest shout is very near the noise level, or when the vector resultant is ambiguous.

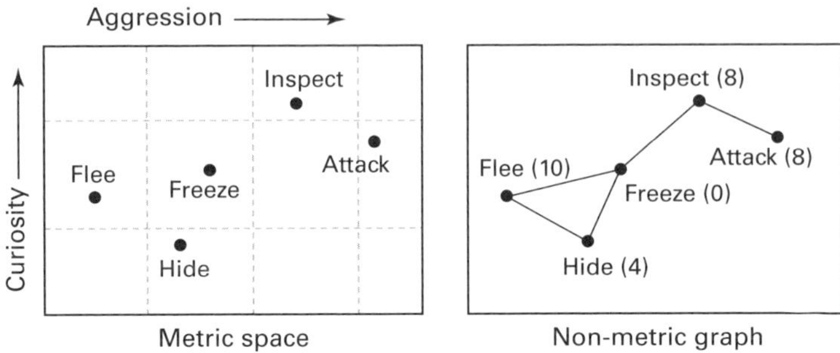

Figure 0.3

A metric space for the behaviors used in figure 0.2 (left panel) and a graphical rendition (right).

2.0   Intrinsic Knowledge

Returning again to figure 0.2, we note that the various actions are not independent. Flee is the opposite of Attack, and hence these actions would be negatively correlated in any rational world. Similarly, inspecting a novel event will require approaching an object, and may be a precursor to an Attack. Hiding has some features in common with standing completely still (Freeze). Such correlations represent intrinsic knowledge about the how behaviors are seen to be related. They can be used to place choices for action in a metric space (Shepard 1962), and provide important constraints for achieving an optimal aggregation of choices, regardless of the complexity of the social system.

To see the impact of shared intrinsic knowledge on group decision-making, let the space of possible actions be related as shown in figure 0.3. These relationships are exhibited in two ways, one metric, the other, nonmetric. Note that in this particular case, the nonmetric graph has been formed by making explicit the steps between regions highlighted in the left panel.

We will return later to the effects of resolution within the metric space on collective outcomes. For now, the focus will be on nonmetric, graphical representations for the similarity relations between choices for actions.

3.0   Social Connections: Bartering

With social contracts in place, and with a graph showing shared intrinsic similarity relations among choices, the possible decision choices for daemons are not independent. Thus, there is no need for one daemon to communicate its desires to another. Furthermore, the shared intrinsic structure provides information that has value for choosing a winner closest to one that maximizes the preferences of the group. For example, what if a daemon's desire could be at least partially satisfied with an alternative choice that is close to, but slightly different from, its first choice? Certainly obtaining this second option would be better than not winning anything at all.

3.1   Preference Orders

Returning to our five daemons, consider again the choices. As implied in the right panel of figure 0.3, linked nodes in the graph share common features. Attack (A) and Inspect (I) are forward movements of the vehicle, the first being more aggressive than the second. Flee (F) and Hide (H) are backward movements, and share the feature of "fear." Freeze (Z), on the other hand, suggests uncertainty between aggression and fear. Given these relations, we would expect daemon F, whose first choice is fleeing, would lend support to either Hide or Freeze over the actions Inspect or Attack that take the vehicle closer to a perceived threat. The reverse, however, will be true for daemon A, which favors an attack and will most likely regard

Inspect (I) as the best backup maneuver. The similarity plots of figure 0.3 make explicit the rank ordering of the alternate choices for any daemon.

We now can engage in a modification of our original social contract, where daemons can have the hope of at least getting a second choice preference if their first choice is not likely. Let us add to a daemon's shout the voices of all its neighbors on the similarity graph (e.g., figure 0.3). These neighbors correspond to a daemon's second choice. We now have included more information about preferences in the decision-making process. For our example, daemon A's level of shouting will be augmented from 7 to $7 + 8 = 15$. Similarly, daemon Z's new shout increases his voting strength from 0 to $0 + 8 + 10 + 4 = 22$. The row labeled "Top Two" in table 0.1 shows that this procedure picks Freeze (Z) as the winner. Note that this winner, unlike those before, is the first or second choice of all but one of the daemons.

A further improvement of social aggregation methods was also suggested many years ago by Borda (1781). The advance was to place less weight on second choice, still less on the third, and so forth. If we restrict ourselves to considering only first and second choices, we would then give a weight of "1" to a first choice, and "1/2" to a second choice, as shown by the rank vector in the second column of the last row in table 0.1. The Borda winner is Flee (F).

By now it should be obvious that the procedure used to aggregate the desires of our daemons can have a huge impact on outcomes (Arrow 1963; Saari 1994). Even if information about the choice domain is incorporated into choosing winners, our daemons may still argue over just how second, third, etc., choices should be weighted when votes are tallied. All will agree that second choices should be included in the count, for then there is a clear individual benefit for the majority. But how do we settle the rank weighting of second choices?

Table 0.1
First tally results

Contract	RankWgts	F(10)	H(4)	Z(0)	I(8)	A(7)	Winner
Vectors	[ ]		*			*	**A**
Plurality	[1, 0, 0]	10	4	0	8	7	**F**
Top Two	[1, 1, 0]	14	14	22	15	15	**Z**
Borda~	[1, .5, 0]	12	9	11	11.5	11	**F**

Note: the proper Borda method is to multiply node weights by their ranks, picking the minimum sum as the winner. The abbreviated reversed Borda used here is more convenient for illustrating anigraf properties.

3.2 The Condorcet Contract

In 1785, the Marquis de Condorcet proposed a scheme that avoided placing weights on lower-ranked preferences. His trick was to compare two alternatives at a time—like a tournament—to determine which one is preferred over the other. No weighting vectors on ranks would need to be imposed; each daemon would simply vote for the alternative in the pair that is more desirable—that is, whichever member of the pair is higher ranked in its preference ordering. If one alternative is then found to beat all others in such a pairwise contest, that alternative is seen as the "fair" social choice for the winner. More important, this Condorcet winner can be shown to be the maximum likelihood estimation for a social order on choices (Young 1995).

Table 0.2 sets up a portion of this tournament. The first row gives the voting strengths of the five daemons. The next four rows illustrate how each pairwise vote is taken. For example, in the first of these rows, F is pitted against I. How will daemon F vote? Obviously, it will choose F over I. Hence, alternative F will receive a vote of +10 from daemon F, as shown in the

Table 0.2
Pairwise Condorcet tally

Pairs	F(10)	H(4)	Z(0)	I(8)	A(7)	Total	Winner
FvsI	10	4	*	-8	-7	-1	I > F
HvsI	10	4	*	-8	-7	-1	I > H
ZvsI	10	4	0	-8	-7	-1	I > Z
IvsA	10	4	0	8	-7	+15	I > A

* = "indifferent" between the two choices

second column of the second row. Next, when daemon H votes between F and I, because alternative F is nearer in the similarity graph (figure 0.3), daemon H will cast its vote for F, adding another +4. Daemon Z's position, however, lies equidistant from both F and I, and hence he is indifferent about the two choices, not contributing a vote to either. Daemon I, of course, votes −8 for itself, the negative sign indicating that the vote is cast for the second member of the pair being considered. Finally, for daemon A, the choice I is closer to its main desire than choice F; hence its vote is cast for I, as shown by the −7 entry. The sum of these entries is −1, indicating that I is the pairwise winner over F. Following this procedure, the winner for the Condorcet contract is shown to be daemon I (Inspect), which beats all the other choices.

The Condorcet pairwise winner thus has three advantages: first, there is no need for a rank vector when aggregating lower-ranked preferences; second, this winner cannot be beaten by any counterproposal, as long as the similarity relationships among the alternatives and weights remain the same; and third, as mentioned, this procedure gives the maximum likelihood choice. Biological systems enforce optimal choices. Thus, although perhaps superficially complex, our social tallies that follow will use this method for aggregating votes, regardless of the complexity of the system. A simple Condorcet tally machine is described in the chapter "Anigraf4".

4.0 Anigraf Abstraction

The mental organisms we have called daemons form a social structure, which is represented by a graph showing the similarity relations between the desires of each daemon. This graphical depiction is called an *anigraf*. Figure 0.3 is one illustration; figure 0.4 another. Note that each vertex of the anigraf corresponds to a mental organism or daemon, with a unique set of preferences. The edges of the graph show which daemons share similar preferences. The anigraf not only makes explicit the similarity relationships among these daemons, but also shows viable communication channels between them. This property follows from the simple fact that meaningful communication of preferences is not possible unless the two partics share in part the same belief structures. The anigraf is thus a social network, but one in the internal cognitive world of mental organisms.

In a fully connected graph, all daemons will share their desires for the social system with each another. In a ring or chain, preferences are shared only with, at most, two other agents. In more complex anigrafs, the connectivity will be extensive, with preferences of many daemons similar to one another. The variety of graphical forms we will study range from simple rings of five nodes to very complex scale-free graphs with hundreds of nodes, each representing a different mental organism with different desires. There will be a strong coupling between intrinsic knowledge captured by a graph, and an anigraf's behavior.

Just as each daemon could control some aspect of a vehicle, so will our daemons be capable of initiating actions, provided

Figure 0.4

Daemons exposed that are controlling movements of a puppet, with the anigraf cognitive architecture depicted as a cast shadow.

the aggregate system makes a collective decision to do so. In the early stages of anigraf evolution, we allow certain daemons to initiate "leg-like" movements or "waves of an appendage." For example, for a chain network, if the aggregate approves the wishes of one of the daemons situated at the end of the anigraf chain, then this daemon might command a traveling wave that propagates through the actual physical embodiment. Such potentials for actions lead us to behaviors like walking or swimming. It is important, however, not to confuse the anigraf, which is a mental construct, with the physical acts that the mental agents initiate. The duality between mental acts and physical actions is best thought of as a set of puppeteers each of whom controls specific physical components of the entity they occupy. If you will, the daemons pull the strings of the puppet in which they reside, thereby translating mental acts into physical actions. Thus, there is a close coupling between decision-making and mechanisms for action. However, the thrust of anigrafs is not the physical structure, but rather the cognitive architectures, and how their designs constrain thought, decision, and action—and consequently behavior.

Part II

Animacy: Action Agents

Periodic, cyclic behaviors are universal among all living systems. At short time scales, cells pulsate, organisms breathe and palpitate, hearts beat, creatures walk, swim, or fly. Classical cybernetics regards these activities as oscillators with feedback control mechanisms. In the more complex systems, there is typically a hierarchy of control. Cognitive capabilities are generally assumed to emerge from this complexity.

Anigrafs explore another possibility: the control of periodic actions is exercised via rudimentary mental organisms called agents or daemons. Each agent is associated with its own mechanism that initiates a particular act or behavior. Although these acts may be reflexive, like a knee-jerk, we may infer there are mischievous daemons that activate them. This set of agents thus represents a distinct level of description, being part of a less confined, social control network that decides just how the system should behave. For this abstraction to be plausible, there must be a competition for control, without a guaranteed unique winner. To create the needed competition, we use the pairwise Condorcet method for aggregating choices. The next few sections explain that, in the presence of such Condorcet competition, periodic behaviors are possible even for very simple organisms. Such "social control mechanisms" distinguish themselves by not requiring classical cybernetic governors.

Anigraf1

Simple Precursors

1.0 Precursors: Two Prototypes

Figure 1.1 shows two prototype anigrafs. One of them, a pentagon, provides a clear but simple example of periodic behaviors. In both prototypes, each node represents one of the mischievous daemons or agents that controls the state of its special physical actuator. Each actuator initiates an action or behavior that moves the system closer to the goal the agent desires. Although one might attempt to capture these same relationships as parts of each actuator, such a model would confuse two levels of description we wish to separate: the level of the physical mechanism versus the level of the relations between these mechanisms.

Our two anigraf examples in the left panel each have the same five goal states (P, Q, R, S, and T), with their associated agents and weights. Each graph shows how the goal states are seen as related by the agent of the organism under consideration. The right anigraf (female) uses a simple pentagon that relates alternatives; the male organism relates the same five behavioral acts as a triangle with a tail. Given the identical desires of both sets of agents, we can proceed as before (see table 0.2) to conduct a Condorcet tally in order to determine which agent's needs are to be satisfied. For the male,

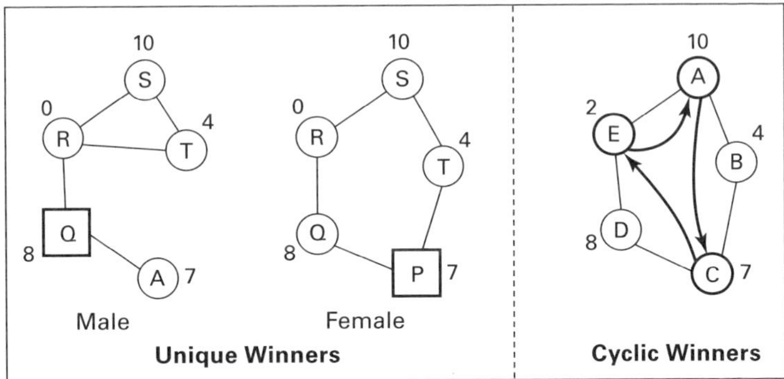

Figure 1.1

Left panel: male and female anigraf; right panel: top-cycle.

it is Q; for the female, P is the social choice. We have, then, two little organisms whose actions are determined by an internal, social decision-making process. From the perspective of the anigraf, the agents or mental organisms controlling behaviors, which range from simple neural modules to complex societies, are represented in a similar manner.

1.1   Dynamics: Top Cycle Definition and Example

Can our small society of anigraf agents exhibit periodic activities, without invoking classical feedback control mechanisms? If so, then this kind of cyclic activity should occur without any changes in the strengths of the agent's desires, but rather be a property of the state and structure of the social network. Obviously, if the aggregation of the agents' whims always yields a unique winner, then only one action ensues. Therefore, tally procedures, such as Plurality, Top Two, or Borda (described earlier, in table 0.1) are not appropriate. However, if agents make pairwise comparisons, then cycles among choices are possible. Specifically, selected weights on anigraf nodes can

Table 1.1
Pairwise Condorcet Tally (Pentagon)

Pairs	A(10)	B(4)	C(7)	D(8)	E(2)	Total	Winner
A vs. C	10	0	−7	−8	2	−3	C > A
B vs. C	10	4	−7	−8	0	−1	C > B
C vs. D	0	4	7	−8	−2	+1	C > D
C vs. E	−10	4	7	0	−2	−1	E > C
A vs. B	10	−4	−7	0	+2	+1	A > B
A vs. D	10	4	−7	−8	2	−1	A > D
A vs. E	10	4	0	−8	−2	+4	A > E
B vs. D	10	4	0	−8	−2	+4	B > D
B vs. E	0	4	7	−8	−2	+1	B > E
D vs. E	−10	0	7	8	−2	+3	D > E

trigger a sequence of actions, without any change in these weights. Such a dynamic allows anigrafs to emerge from simple cause-effect, feedback-type reactions to behaviors with a more choreographed repertoire of actions.

The pentagonal anigraf at the right of figure 1.1 illustrates how cycles in outcomes can emerge. This anigraf, A . . . E, is identical to the adjacent "female" pentagonal anigraf, P . . . T. The only difference is a very small change in one of the input weights: agent E has weight 2, rather than the 0 weight held by agent R. Because we might expect the small weight change to have little consequence, the likely winner in anigraf A . . . E is node C, which is comparable to the previous winning node P in the female anigraf, P . . . T. Indeed, as shown in the first rows of table 1.1, C does beat A, B, and D. However, note that C now fails to beat alternative E (fourth row). Therefore, we further examine the remaining pairs and find that not only will A beat E, but B and D will also beat E. Each time an agent dominates the cycle, the agent effects a new behavior of the vehicle that embodies it. Thus, we have at least a three-cycle among pairwise comparisons: A beats E, which in turn beats C, which now beats A. These dynamics continue until a new set of weights is introduced. We have created what is called a *top-cycle* in the choice of behaviors.

Definition: A top-cycle among alternative choices occurs when there is an alternative A_i that beats A_j, A_j beats A_k, and A_k beats A_i, and every alternative not in the top-cycle is beaten by at least one alternative in the top-cycle.

Note that a Condorcet tally must lead to a top-cycle if no alternative beats all remaining alternatives in the pairwise comparison (excepting ties). Furthermore, if there is a top-cycle in the outcome, then there must be at least a three-cycle, such as in the previous example. Top-cycles add a dynamic that plays an important role in the behaviors that follow.

1.2   From Cycles to Chaos

Cycles are typically a prelude to chaos. There are two obvious ways to create chaotic sequences of outcomes. The first, and more difficult, is to choose appropriate weights on special graphs like the pentagon, so that the dominant choice is overturned in a pair-wise double check. For example, in table 1.1 C is overturned by E. A second, much simpler method is to allow agents to vote haphazardly, ignoring the relations among choices specified by their anigraf. In this second method, as the social system becomes larger and larger, cyclic outcomes become more and more likely, with 1 as the asymptote. The probability of top-cycles is already two-thirds for six-agent systems; and for a group of twelve agents, 90 percent of the weights on nodes that reflect agent desires will result in a top-cycle among the choices (see figure 1.2, squares). If these

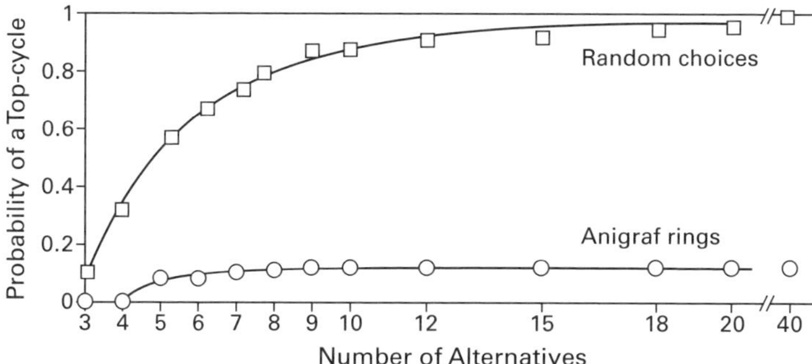

Figure 1.2

Probability of top cycles for anigraf rings for an increasing the number of alternatives, each having random weights. Also shown is the top-cycle probability for Erdos-Renyi random graphs.

Box 1.1
Why Chaos?

To provide some insight into why chaos, we need a more detailed analysis of the relationship between the preference orderings and the calculation of pairwise winners. The functions of interest are an expansion of the set of pairwise outcomes $[_nC_2]$ in the n unknown weights for the agents. If each pairwise comparison is independent of all others, as it would be for random weight choices, then there will be $[_nC_2]$ equations in n unknowns (the weights of n agents). The random assignment of weight to each row provides ample opportunity for finding at least one case in which any agent will be beaten by another in the pairwise comparisons. Therefore, there will be no Condorcet winner, and cycles in outcomes emerge (Saari 1994).

weights are then perturbed slightly from tally to tally, the result is a chaotic sequence of outcomes.

To reduce the likelihood of cycles, agents' preference rankings must respect the anigraf form. But this in itself is not sufficient. As illustrated earlier with the pentagon, figure 1.2

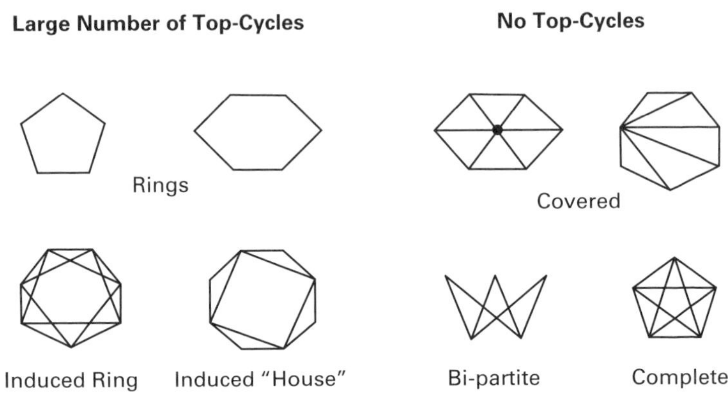

Figure 1.3

Simple anigrafs that will have top-cycles for some set of weights on nodes, and others that will not, regardless of the node weights (Richards, McKay, and Richards 1998 and 2002).

shows that all ring anigrafs can have top-cycles, and these will occur for about 12 percent of the weights on nodes that reflect the desires of the agents—assuming these weights are chosen from a uniform distribution. Other anigraf forms that can lead to top-cycles are shown in figure 1.3. Furthermore, whenever these forms appear as components of an anigraf (e.g., an induced ring, or induced "house"—a pentagon with one extra edge), top-cycles in outcomes may occur—although the probability of encountering a top-cycle is typically much less than for a ring anigraf.

One might now wonder whether there are graphical forms guaranteed never to have cycles. The answer is YES. If the anigraf network is completely connected or is "covered" (i.e., one node is adjacent to all others), or is equivalent to a complete bipartite graph, then there will be no Condorcet top cycles given that each agent's preference orderings are consistent with the graph structure. Other examples are shown in the right panel of figure 1.3. These "stable" anigraf forms can be contrasted with those having a high probability of top-cycles, as shown in the left panel. Note the addition or deletion of a single link in the network can cause significant changes in whether or not the social system can easily reach a unique consensus. The connectivity of the network thus also plays a key role in the potential for cyclic outcomes.

1.3   An Example

The space of winners and losers (top-cycles) for an anigraf will be of dimension n, where n is the number of alternatives. Regions within this space will be divided into polyhedra, with the boundaries corresponding to a change in an inequality of weights on alternatives. Figure 1.4 shows a slice through the five-dimensional space of a house graph relating the alternatives A, B, C, D, and E. Three of the weights are fixed, and the remaining two range from 0 to 10, as noted in the caption to the phase plot. Zero values for these two variable weights (x, y) lie at the upper-left corner of figure 1.4.

The first striking property of any phase plot is that the boundaries between regions are straight lines. These boundaries occur where an inequality between two weights is reversed. The transition from A (red) to B (yellow-green) is one example. Similarly, the transition from B (yellow-green) to the textured region of top-cycles indicates another change of weight inequalities, as do the boundary between the top-cycle region and the green polygon at the right. These changes can be used to infer the members of the top-cycle.

Figure 1.5 shows a horizontal cut through the phase plot at $y = 6$, which is a line that grazes the top of the "x" junction of yellow-green, purple, and the two adjacent top-cycle regions. As we move along this line from the left edge of the plot, increasing the weight on node C incrementally, we begin with

Figure 1.4

Phase plot showing winners (single color which indicates node) and top-cycles (textured regions) for a house anigraf with weights (6, 3, x = 0–10, y = 0–10, 1). The five vertices are labeled clockwise beginning with tip of roof.

a "winner" region of purple (node E), encounter a slight bump when entering a small yellow region (node 2) before the appearance of the top-cycle C > D > E > C . . . 3 > 4 > 5 > C, etc. Then, as the weight of C is increased to 7, we move into the stable green region at right.

1.4   Summary

Anigrafs have a wide range of potential behaviors, ranging from unique, stable outcomes (87%) to chaos when the anigraf model relating alternatives is ignored. For stable outcomes, an agent or mental organism's ranking of choices must be

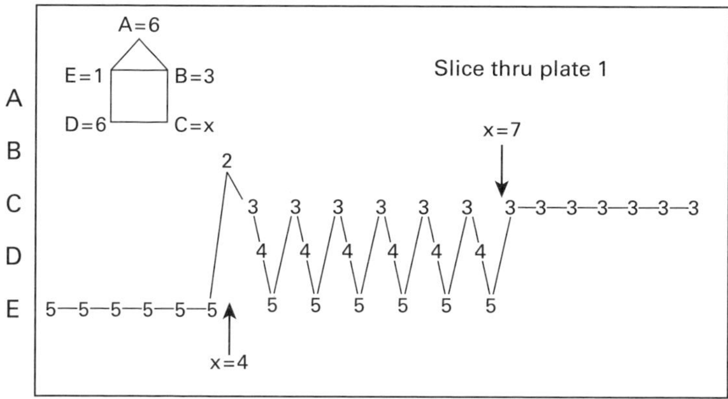

Figure 1.5

As the value of x is increased, a top-cycle (3, 4, 5) replaces winner 5, which turns into a new winner 3 after six iterations (i.e., "time steps").

consistent with the anigraf form that relates these choices. This form represents the shared, or common, intrinsic knowledge about the choice domain. Of special interest is that about 12 percent of the time, nonchaotic cyclic outcomes can happen for certain choices of weights, even when the anigraf form is respected and neighbor-only communications are in effect. In these cases, the tally machine will deliver a cyclic sequence of outcomes until new weights are entered. Even with random weight selections for agent desires, this behavior occurs with a level of significance that provides some simple anigrafs, such as rings—and later, also chains—with the potential to achieve pulsating or rhythmic activity typical of very simple animate cells. In contrast with classical control systems, in this case it is not a physical feedback mechanism that underlies the cyclic sequence, but rather the manner in which information is shared: who talks to whom about the preferred next state for the system as a whole.

Anigraf2
Swimmers: Beginning to Move

2.0 Movement

Precursor anigrafs had very limited behaviors, constrained for didactic purposes to simple ring forms. This simplification hid the potential complexity of designs and behaviors that could be attributed to primitive life forms. Here we begin to expand this repertoire. The component mental organisms, or "agents," will have access to different kinds of interfaces with the environment, with interface hardware that affects the behavior of the system as a whole. Each agent will have control over those additions, and will "vote" depending upon the strength of its desire to achieve a preferred goal. In effect, then, we are formalizing a two-tier system. On one tier, the interface agent constitutes a member of the social system of all other agents; and on another, lower-level tier, each agent controls an effector, but only when approved by all other agents.

Among the most visible and prevalent effectors are flagella, or hairy muscular fibers. For example, the single-cell Euglena has a flagellum that waves to pull the creature forward; another simple creature, the paramecium, has a host of hairy superficial fibers that can generate wavelike motions. These appendages can be entirely passive, with the movements initiated only at the point of attachment to the cellular body. More complicated

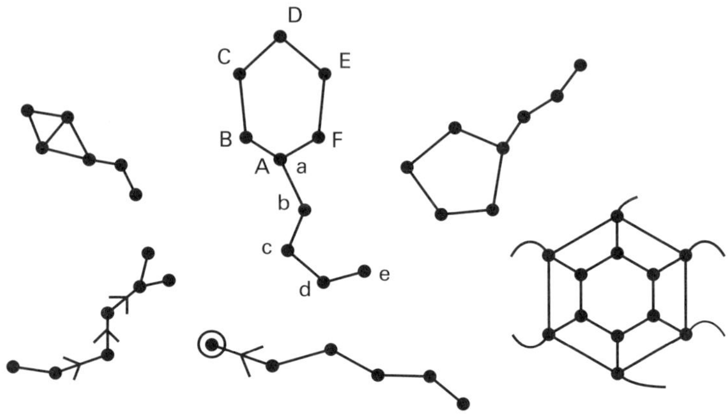

Figure 2.1

Potential swimmers, some with flagella.

limbs may have additional agents confined to "local" social enterprises that control special sinusoidal motions, such as the movement of wings, fins, legs, and eventually fingers. We begin first, however, with a description of a society of agents that can generate chaotic flagellations typical of coelenterates. This little society has a simple objective: to move one way or another.

2.1   Jellyfish

Coelenterates have very primitive neural nets that lie on the rim of a funnel-like body. When activated, this network causes a wavelike motion of tentacles attached to nodes on the rim. The jellyfish is an example. It has an inverted, cuplike form with many tentacles emerging from the rim of the cup. Typically the neural elements embedded in the ring appear in multiples of four. Each neuron drives its own tentacle, causing it to flagellate. The anigraf analog is thus a group of agents linked together in a ringlike configuration, such as illustrated in figure

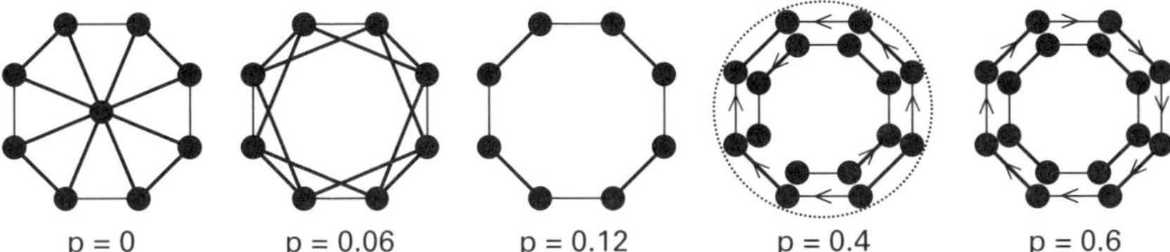

Figure 2.2

Ring-like swimmers with their top-cycle probabilities.

2.2. Of primary interest is how likely this type of anigraf is to generate cycles among actions initiated by its constituent agents.

When the anigraf form is a simple ring, then each agent communicates only with its two neighbors. In figure 1.2, the probability of cycles was shown to reach a peak at 8 to 12 agents. Let us now consider design possibilities for an octagonal jellyfish anigraf, recognizing that both cyclic and noncyclic outcomes are useful. An example of a cyclic outcome would be a top-cycle that moved from one agent to another, rotating around the ring, and hence initiating a circular wave of activity. A noncyclic outcome, on the other hand, might result in the coordinated constriction of the rim of the coelenterate as it captures a food particle.

Figure 2.2 shows a few modifications of the simple ring, and the probability of top-cycles among outcomes using the Condorcet tally, where the weights on nodes (i.e., the strength of the agent's desires) are chosen from a uniform distribution. If the ring is reconfigured to a "wheel," then the anigraf has one vertex that is adjacent to all others (i.e., the graph is "covered") and there will never be cycles ($p = 0$.) The next configuration is a regular graph, where each node has a degree of four. The top-cycle probability is about 6 percent, which is roughly half the cycle probability of the simple ring. Unfortunately, none of

these percentages may be high enough to ensure that our jellyfish can easily engage in complex, long-cycle flagellated
movements.

A simple modification of our communication channels
solves this problem: we make the communication channels
directional. In other words, let preferences be "passed on"
from one agent to the neighboring agent, but not vice versa.
Depending upon the directionality of the communication channel, we also will have the great advantage of being able to
enforce either clockwise or counterclockwise cycles among the
rim agents. For the jellyfish with a ring of eight agents, clockwise communication will produce clockwise cycles 60 percent
of the time; for a ring of twelve agents, the cycle probability
rises to 80 percent. These numbers are approximated very
closely for cases in which $n > 4$ by cycle percentages shown
earlier (in figure 1.5) for random preference orderings. (See
also figure 2.3.) The difference, however, is that now all agents
respect the shared (directional) anigraf model when ranking
their preferences. Clearly we have the beginnings of a social
control system that mimics a physical feedback controller, but
in this case the condition of sharing social preferences is met.

A periodic wave of tentacle movements is only one of several actions that the jellyfish might engage in. Even if the agents
all have identical effectors, all effectors need not perform identical movements, and, consequently, all agents need not have
identical goals for the system. Some agents might prefer that
the jellyfish move to another, more favorable region in its environment (more "food" or perhaps "less hostile"). Another
might want to ingest nutrients. Yet another might wish to
release a toxin. As in our primitive single-cell anigraf1, some of
these actions might entail a change in shape—perhaps to be
carried out by activating other internal fibers or membrane
properties. In a very simple scenario, there could be two types

of agents on the coelenterate's ring network, X and Y, where X agents would create an inward current and Y agents would create outward currents, or tighten the membrane supporting the ring. All these choices entail societal consensus that is reached after conducting the Condorcet tally among the agents. This coelenterate ring of agents can easily be regarded as a very primitive brain.

2.2   The Flagellum

Consider next yet another primitive creature, such as the sperm cell, or Euglena, where a single flagellum controls movement of a simple, cell-like "body." If this appendage itself is passive, driven by forces applied to its base, the effect will be as if someone were whipping or twirling a cord. Such a cord will have some kind of natural frequency of motion when torque is applied at one end. As shown by Berg (1996), these simple creatures have a micron-sized motor that whips the miniature tail. Let's say agent A is responsible for initiating movement of this flagellum, thus satisfying an exploratory need to move forward. We know from table 1.3 that if voting strengths of the agents are suitably chosen, cyclic outcomes will occur that will include agent A's preference for system behavior. The little creature continues to flagellate until new voting strengths are tallied.

A variety of cycles that include A's preferences are possible for our Euglena. Consider both hexagonal and pentagonal anigraf forms, as illustrated in figure 2.3. Symmetric top-cycle activation of agents seems most likely for forward movements, whereas asymmetric cycles offer an option for changing the direction. We can also distinguish top-cycles that include the neighbors of A, as in the pentagon, or, alternately, cycles that exclude these neighbors, as in the leftmost hexagon. Each of

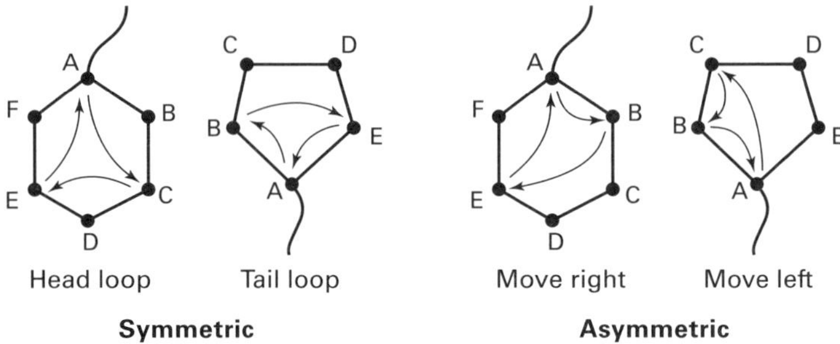

Figure 2.3

Euglena-like anigrafs, showing possible cyclic movements.

these top-cycles has its advantages. For example, switching to other cyclic states may be easier if two members of the symmetric top-cycle are nonadjacent to A. On the other hand, when neighbors of A are part of the top-cycle, then agents such as C and D in the pentagon, or C and E in the hexagon, are free to reinforce each other's goals. An additional important benefit of including agents nearest to A is that there is a tenfold greater possibility of top-cycle activity. Forward or backward movements are thus the most likely. Clearly, even for our little Euglena anigraf, many different movements seem possible, and these different movements are capable of supporting a variety of behavioral goals.

2.3 "Smart" Tails

To create creatures with more complex locomotion abilities, we may wish to control the amplitude or form of the wave of the tail, or bias its body angle to change direction. These effects cannot be accomplished easily by a single, simple agent at the point of attachment of a passive appendage. Let us then break the tail into segments, simultaneously creating a chain of

Smart Tail: 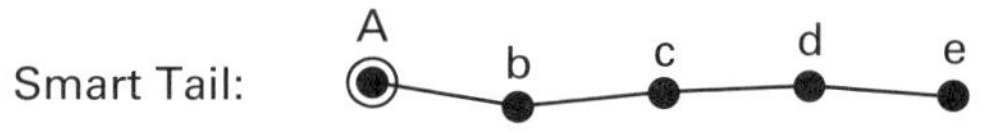

Figure 2.4

A simple worm-like creature with A its "head" controlling a wave.

agents, each of which controls the activation of one segment. Let one of these agents, say A, be selected as the "head," using uppercase to make this designation (figure 2.4).

Because a chain is just a broken ring, the top-cycle behavior of chains is very similar to that of rings. Our objective is to create a cycle of activity through at least one part of the A, b . . . e chain anigraf. In turn, this cycle among agents will effect a physical movement of the tail or body, such as when a fish swims or a snake crawls. We assume that the agents are embedded within a flexible shell that contains contractile tissue such that when an agent is active, muscular springs will contract to create part of a wave motion. A cycle among agents—say, agents A, b, c or A, b, d—can drive this kind of behavior.

Not surprisingly, if there is no sharing of information among agents, then there will be no cyclic behavior. Intuitively, if all agents are independent, then in a pairwise competition among choices, the agent with maximum clout will win. Similarly, although less obvious, if all agents share their current preferences with each other, placing a preference when information is shared only among immediate neighbors, only the top two alternatives in an agent's preference order are used, with all remaining alternatives (at level three or deeper) having weights reassigned as zero. The number of levels used is called "knowledge depth" noted Kd (see figure 2.5). We will typically use Kd = 2, which will allow cycles even for short-chain anigrafs.

Definition: Knowledge depth (Kd) is the number of levels in a preference order in which the weights are typically nonzero.

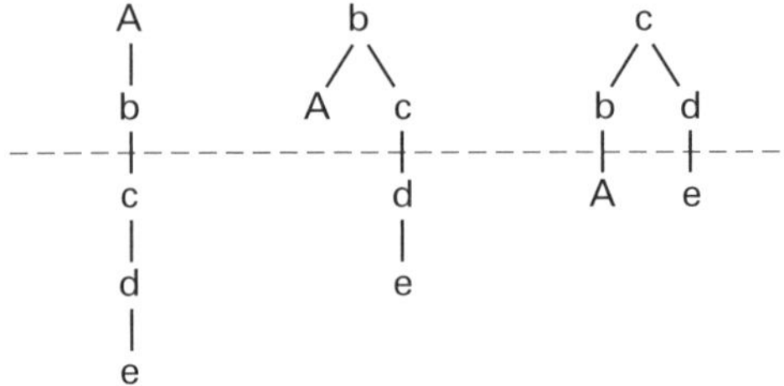

Figure 2.5

"Smart tail" example of full knowledge depth (with knowledge depth 2 shown above the dashed line).

For lower levels, all alternatives are assigned zero weights when calculating the strength of an agent's wishes. Full knowledge depth uses all the weights in a preference order.

A second way to obtain top-cycles is to use anigrafs with directed edges. Figure 2.6 and figure 2.7 illustrate. With undirected communications between neighboring agents, the top-cycle probability reaches a maximum of about 15 percent for a chain of ten to twelve agents (diamonds). If directionality is introduced into the social network (plusses), then cycle probability rises five times (to between 60 and 70 percent), increasing further as the chain length increases, and eventually reaching 1 as the asymptote. Directed ring anigrafs behave similarly (circles). Anigraf social networks composed of chains thus have a high potential for eliciting periodic activities. Even rather small, eight-chain anigraf "worms" can initiate five-cycle waves, such as AbdecA, with directional communications.

2.4   Cooperative versus Competitive Networks

Anigrafs with directed, rather than bidirectional, communication channels come at a cost. Such networks marginalize one of

Figure 2.6

Phase plot showing winners and top-cycles for a directed chain ani-graf with the weights x = 0–10, y = 0–10, 2, 6, and Kd = 2. Textured areas are top-cycles.

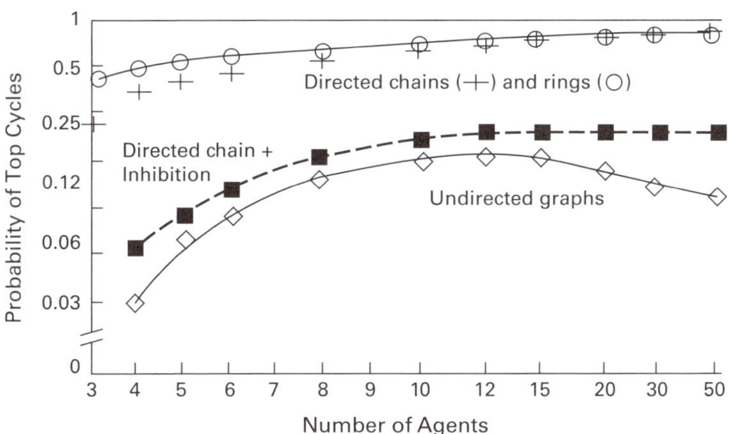

Figure 2.7

Effect of inhibition on directed graphs.

our key assumptions about social awareness. Because information is passed only one way, and not back and forth, the sharing of knowledge and preferences within the system is quite limited. Such networks resemble those with feedback loops. As a species, anigraf creatures with directed networks thus are second-class citizens, and lie between a fully aware (bidirectional) anigraf and Braitenberg's more reflexive vehicles.

To clarify this connection, consider how we might shut down or control top-cycle activity, especially if it leads to chaos. Rather than removing directionality, a simpler solution is to shut down the voting weights of the agents (or, equivalently, assign them equal weights). This is easy to implement through either global inhibition or inhibitory feedback loops among agents. As the exchange of information among agents is reduced to zero, the probability of top-cycles decreases and can be driven *below* the bidirectional result, as shown in figure 2.6. (The inhibitory case plotted shows the attenuation effects, but not the limiting result.) The more agents that have clouts reduced to zero, the greater the chance of one agent getting its way. Our directed anigraf has then been morphed into a kind of hybrid between a Braitenberg vehicle and the truly social, bidirectional anigraf.

Anigraf3

Walkers: Syncopated Limbs

3.0   Enhancing Behaviors

Swimmers have limited freedom of movement. Their actions are thus limited, and, consequently, we do not expect such creatures to possess a high level of intelligence. Our most advanced swimmers use wavelike motions of a segmented appendage for locomotion. Resident within each segment is an individual agent. A chain of these agents guides a wave motion of the body or limb segments. To enhance the behavioral repertoire of such anigrafs, an obvious next step is to add more limbs. Alternatively, the chain of agents itself can be augmented, such as adding branches so the anigraf resembles a simple tree. In each case, such additions place significant demands on the control structure of the anigraf "brain" required to coordinate the various agent activities.

To illustrate these hurdles, we follow an evolutionary path, and begin with creatures having many legs. As before, each leg has its own set of low-level agents that govern the type of movement of the limb. Movements of these separate appendages, or, more specifically, the activity of these independent sets of low-level agents, must then be coordinated to create a sequence of limb movements. The pattern of these sequences is a gait.

To solve the difficult problem of retaining body stability under gravity, we begin with six or more agent-controlled limbs, or legs. The most common gait is to lift successive pairs of limbs in a wavelike motion, thus insuring that most of the legs continue to support the body. This is a first step in an exploration of whether realistic gait patterns can be created in a single chain of body agents, or whether a control structure having two linked chains, such as a spinal cord, is more plausible.

3.1   Centipede Anigrafs

Imagine a creature with many limbs on each side of an elongated, cigar-shaped body—such as the sixteen-legged anigraf illustrated in figure 3.1. From head to tail, label these pairs of legs {al, ar},{bl, br},{cl, cr} ... {nl, nr}. Because the legs in each pair are moved together, we can place the controlling agents in a central chain, or "cord" (represented by the star symbols in figure 3.1). Then a wavelike motion of the limbs requires an n cycle among the agents, ordered sequentially, as in a cascade. But already we know from anigraf2 that if $n > 6$ (or, certainly, if $n > 8$), a directed chain of communications between

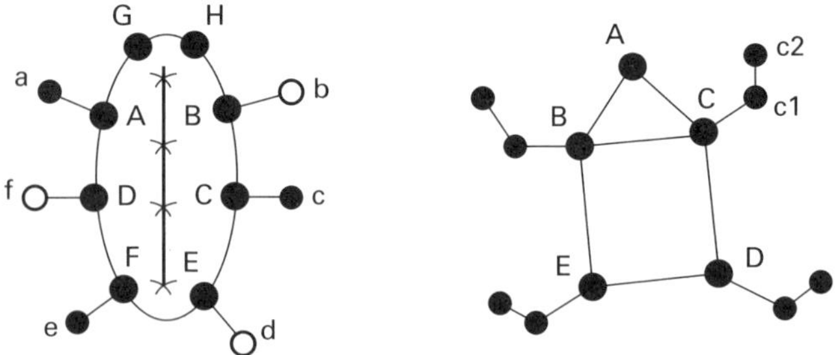

Figure 3.1
Two forms of legged anigraf creatures.

neighboring agents will almost surely be needed. Furthermore, for large *n* only a very restricted set of voting weights will create the desired wave. In effect, the anigraf construction has been reduced to a reflexive machine. Although a small group of agents located at the head of the cord could trigger a wave motion, it is unlikely that any social network could create a cascaded wave that propagates down a cord with many segments. Social networks that control gait patterns must be limited to six or fewer agents. This implies that the creature should have no more than six limbs, or, at most, six states, if pairs or triplets of limbs are linked together.

3.2   Cockroach Anigraf

This six-legged creature has two gaits: one is a wave; the other is called the "tripod." The wave gait is the same as in the centipede, but with fewer legs: first the front two limbs are moved, then the middle two, and finally the last two. The tripod gait, on the other hand, simultaneously moves three limbs located at asymmetric positions, as illustrated in figure 3.1 by the solid and open nodes. Both schemes preserve balance.

As we have seen with the centipede, either the cockroach wave gait or the tripod gait could be produced by some simpler reflexive automaton, activated when a social network or "head" agents pick either agent T (tripod) or agent W (wave) as the winner. On the other hand, unlike the centipede, the cockroach has only six legs. Now only four agents are needed along a central cord to create the wave gait. Furthermore, many different top-cycles can be created. This opens the door to a socially based anigraf network for innervating limb motions. We can move still further away from automata-like systems by positioning one more agents on each of the six limbs (see figure 3.1). These "second-tier" agents, a–f, can serve two functions: (1) they can provide a basis for invoking local cycles such as B, C, c, D, B; and (2) they have the potential to bend a two-segment limb.

3.2.1 Limb Configurations

Although we could proceed as before to determine the voting strengths needed to generate sequential limb motions, it is more productive to consider first each limb in isolation, driven by its own local cycle. With neighbor-only communications (i.e., Kd = 2), the minimum number of agents needed to create cycles is five, if the agents are networked by a ring or a tree, and four, if they are networked by a chain. Obviously, at least one of the agents must be part of the body, because this will be the agent that is activated by consensus with the other body agents.

Figure 3.2 illustrates some of the minimal arrangements for agents that could govern limb control. In each case, we have incorporated the top agent (large circle) as a member of the anigraf's social repertoire. However, such agents must also be

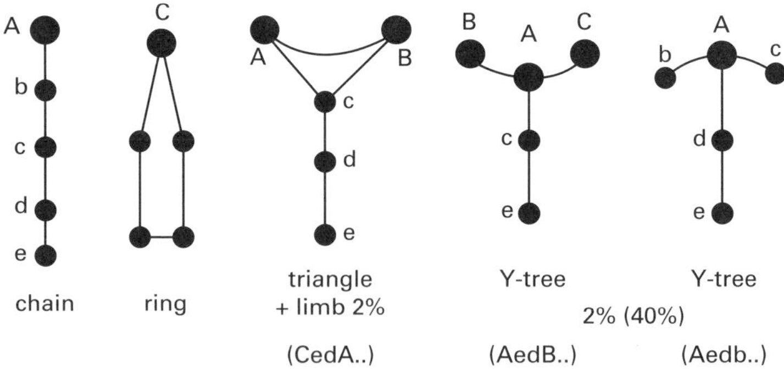

Figure 3.2

Some minimal arrangements for limb control. Large circles represent agents in the "head," outside the limb itself.

linked to others having the same social status. Hence a possible variant is to have two agents, rather than one, as part of the body structure (see the middle illustration in figure 3.2). However, if two agents are on the body, then the limb must have two segments. This follows because a one-segment limb would have a total of only four agents, with one connected (or covering) the remaining three. Therefore, we need a minimum of five agents, which is satisfied with a two-segment limb.

Placing two agents on the body creates a possibly awkward triangle arrangement at the head of the limb. A more agreeable configuration is to have three agents at the top, with two other agents on the limb, thereby forming a Y or a T-tree, as shown in the last two illustrations in figure 3.2. Now all three of the top agents can belong to the body set of agents, or, alternatively, only one of these agents would be part of the body set and the remaining two would be lower-level—actually being part of the limb activation system, which here we represent as an upward arrow configuration. Unfortunately, in all cases, the probability for achieving cycles using random voting is only 1 to 2 percent for undirected graphs. However, with a three-level restriction on weight choices (e.g., mid, hi, lo, lo, hi), the range of levels used in chapter 2 will yield a 40 percent or higher probability of certain four-cycles. Now, with weights "d" and "e" restricted appropriately, either an AedB or and AedC cycle can be activated simply by raising agent A's clout from zero to the upper third of the range of weights.

3.2.2 Spinal Cord Configurations

In animate forms, limb movements are sequenced in part by activity that moves down a spinal cord. A set of stacked Y-trees or T-trees resembles this spinal configuration. In figure 3.3, a three-level cord is illustrated with four sets of limb segment agents (open circles, attached at positions A, B, D, and E). The

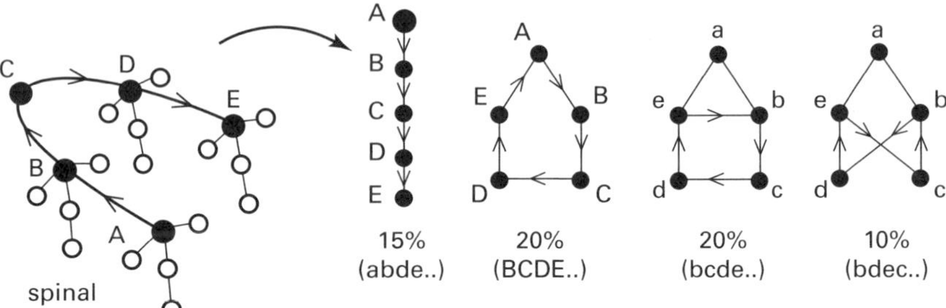

Figure 3.3

The left configuration is a prototype spinal cord. The remaining four graphs show possible sequences and their probabilities for random weights on nodes.

bilateral symmetry of the cord is forced by the need to place limb segments on opposite sides of the anigraf. Activating local cycles in each limb segment now becomes trivial. First we freeze weights for the limb agents, choosing suitable values for cycles. Then we need only activate A, B, D, or E in the cord. A more abstract version of this chain can be obtained by unfolding the cord, as illustrated in the second graph from the left within figure 3.3.

Other, probably less attractive, modifications to the spinal cord configurations would include closing the spinal chain to form a ring, or joining bilateral components of the cord to create a "house" graph. We can also influence the type of cycle desired by changing the graph structure. To generate figure-eight cycles, nodes should be connected in a crisscross manner, as illustrated at the far right of figure 3.3. In each case, the cycle can be turned on or off by appropriate settings of the weight or clout of a body agent. Clearly, directed channels play a key role in setting up the desired cycle. To reiterate, with undirected channels, the odds for a four-cycle are typically only 1 or 2 percent; with directed channels, about 10 to 20 percent of random weight selections will be successful, with still further improvements using three levels of voting strengths. The combination of both undirected and directed channels can ensure that over 80 percent of the voting weights will yield an *abde* cycle for the simple chain. Thus, three factors govern how easily four-cycles can be generated using social networks: the configuration of the graph; the choice of which edges (channels) are directed; and whether voting strength is chosen from uniform distributions, or from one of three ranges (e.g., lo, mid, hi).

3.3   Quadruped Gaits

Generating cyclic movements for four-limbed or two-limbed anigrafs appears plausible using one of the several variations of spinal cord networks described. A remaining issue is to control the sequence dynamically in order to maintain balance, as the transfer of power moves through the cord. Surprisingly, Raibert (1988) has shown that a rather simple machinery suffices, if the limb (and body) motions satisfy certain "sinusoidal" constraints. The problem is then the coordination of the cyclic activities: we must have a clocked sequence and interleaves the limb movements properly.

To engage in a walk, a minimum of a four-cycle among agents is required, even if we have a biped anigraf. (There can be no two-cycles.) For our minimal quadruped to maintain balance, the right front (Rf) and left rear (Lr) limbs should move together, and similarly for the two remaining limbs (Lf, Rr). Therefore, only two sets of limbs need be controlled, together with two appropriate pauses between movements. Alternatively, we might eliminate the pauses altogether and use each agent to activate limbs in sequence. This would generate an Lf, Rf, Rr, Lr sequence.

Table 3.1

Gait Type		Sequence		
Walk	Lf & Rl	pause	Rf & Rr	Pause
Jump	Lf & Rf	pause	Lr & Rr	Pause
Cantor	Lf & Lr	pause	Rf & Rr	Pause
Run	Lf	Rr	Rf	Lr
Bound	Lf	Rf	Lr	Rr
Gallop	Lf	Rf	Rr	Lr

Table 3.1 lists other gaits possible with the simple four-cycles. How can this set of gaits be generated using one social network? Ideally, we would like to envision the top-cycle sequence of agent activities as having a simple mapping to the desired movements. To accomplish this, the anigraf map of the similarity relationships between agents needed to have both left-right and rostral-caudal symmetry, because this is a property of the gait sequence. The pentagon, or "house," with directed communication channels is the simplest anigraf that satisfies this condition. As shown in figure 3.3, paths through such directed graphs correlate nicely with the resultant sequence of the legs. Therefore, unlike most linear networks with resonant modes (Greene 1962), we have a pleasant mapping between how agents view their relationships to one another—a cognitive stance—and the cycles among the limbs affected by the agents—a physical consequence.

Note that the above proposal also requires a two-tier system: a set of (five) body agents, or mental controllers, that specify the sequence of limb movements, and another subset of three or four "robo-agents" that activate the bending movements of each individual limb. These local agents should be seen as part of the lower tier Y or T network, with at least one of the body agents lying at the junction of the network. This body agent is in a key position: its vote can simultaneously either activate, or shut down, both the cycle among the body agents and the local cycle among the robo-agents in the limb. In this case, both the activation of the limb and its own bending movements will be coordinated in the sense that there is one "body" agent in common.

3.4   Gait Switching

One of the great benefits of a social computation is that behavior can be changed dramatically, yet predictably, by one vote. For example, we may wish to change the quadruped's gait from a "walk" to a "gallop," or perhaps simply change a walk from forward to backward. One obvious scheme for gait switching is to change directed channels in the graph, such as from a "house" to a "crisscross" construction. (See figure 3.3.) However, more preferable would be to leave the network unaltered, and simply change the voting power of one or more agents. The use of three-level weight selections makes switching gaits even easier.

A precursor to simple, but sophisticated gait switching is already present in the limb segment. The top of a Y or T network has three agents, at least one of which is a body agent. Cycles involve either the right or left arms of the Y or T, but not both. If the Y configuration is used with three body agents, then each of the two cycles could make slightly different adjustments of the limb position, as would be necessary for either forward of backward movement. Similarly, if a T configuration is used to control limb motion, then the two different local cycles might effect a small lateral adjustment of a limb motion, as would be needed for turning either left or right.

More complex gait changes involve going from a walk to a run, or to a gallop. Referring to table 3.1, a "run" maps into a

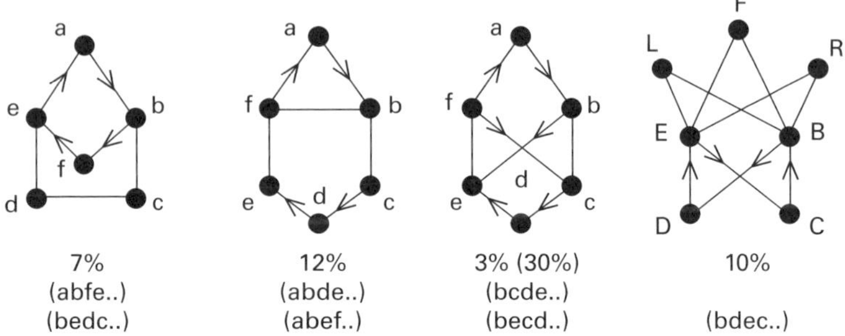

Figure 3.4

The probabilities of cycles when using three levels for weights on each node.

four-cycle around the lower (square) portion of a house (see figure 3.3), whereas the "gallop" requires a crisscross sequence shown in the adjacent graph. To obtain both four-cycles using the same network, and for a significant portion of the weight space, it is convenient to add a sixth agent. One suitable configuration is a hexagon with two directed diameters, as illustrated in figure 3.4. Another is a directed k-partite graph (right). For the modified hexagon, two relevant four-cycles are becd . . . (gallop) and bcde . . . (walk). Only about 3 percent of the weight space will support these four-cycles. However, once again, the use of three range levels of weights will yield cycles 30 percent of the time, or more if the range of each level is narrowed. One suitable set of weights for becd and bcde cycles are (lo,mid,hi, hi, and mid,lo). Note that the difference is simply a switch in the voting strengths for the second and third agents.

Other configurations of interest include the pentagon with a sixth node added to the interior. With channels directed as illustrated in figure 3.4, we can obtain either an abfe or a bedc cycle. The latter corresponds to the customary four-legged (quadruped) walk. The abfe cycle, however, engages only two of the four limbs. Various roles can be assigned to agents A and B. One intriguing possibility if the agents controlled both body stance and balance, as well as a "pause". Then the abfe cycle would allow an "erect" anigraf to engage in a walk, while the bedc cycle for the same anigraf would control a four-legged run. In each case, suitable weights for these cycles are potentially accessible using a three-level voting strength.

Approximately 18 different gaits have been observed in quadrupeds (Hildebrand 1966). With five agents, we have about 300 possible acyclic connected digraphs; with six agents, about 6,000 possibilities; and with eight agents, more than 20 million. Thus, there are many configurations available to generate 18 different gaits. However, it is quite difficult for any given configuration to generate more than a few four-cycles over a significant range of weights. If six or eight cycles are required, then plausible solutions almost vanish. One would have to overlay several different graph structures, and then activate each independently, or allow "online" modifications of the structure of the digraph. These properties are quite unlike linear resonant networks that have many $[O(n^2)]$ solutions for any one four-cycle. Curiously, in real animate creatures, the highest number of gaits observable is about six.

3.5   From Gaits to Goals

For very simple anigrafs—those we regard as akin to the minds of insects or other reflexive creatures—we have a situation not unlike our five daemons trying to control a vehicle. Specific goals such as "flee," "approach," or "avoid" are closely associated with particular locomotive movements such as "run," "walk," or "turn." In cases where there is a simple one-to-one relation between the gait and goal, we can assign specific goal states to each gait agent. To illustrate, let the set of goals "flee,"

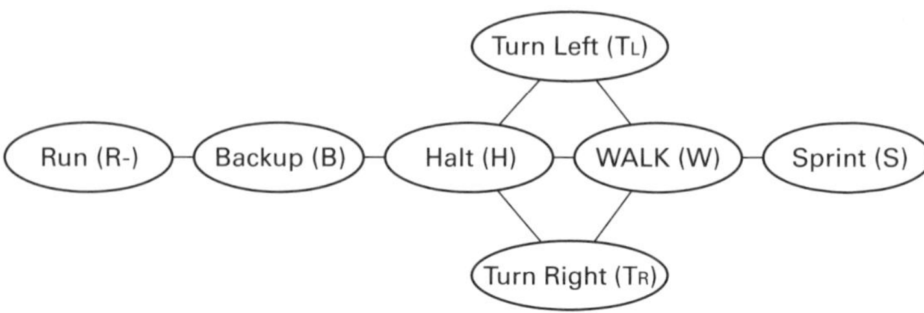

Figure 3.5

Relations among locomotive states.

"attack," "approach," "avoid," "reflect/watch," and "retreat" each be associated with the movements of run away (R), sprint forward (S), walk (W), turn right or left (Tr, Tl), halt (H), and backup (B). How the creature behaves in a given context will then depend on how these movements and their associated goals are related.

One possible relationship for these behavioral acts is the tree anigraf shown in figure 3.5. Fast attack (sprint forward) is seen related to an approach (walk forward), while backing up is a form of retreat, with fleeing the extreme version. Turns are usually executed at low speeds while walking, or when stationary. Although cast in terms of movements, the controlling agents now are engaged in another level of social consciousness, one having a clear emotional content. Flee implies fear, approach implies curiosity, attacking is a form of aggression. These agents have a character one might associate with the daemon-like mental organisms. Cognitive capabilities are emerging.

Anigraf4
Tally Machines

Swimmers and walkers illustrate the anigraf abstraction, being focused on cognitive aspects of knowledge and decision-making. Here we depart briefly from these abstractions to explore how agents' voting might be implemented. For animate creatures, neurons or neural assemblies are the analogs of cognitive agents. Similarly, the tally process will also be a collection of neurons organized to aggregate inputs from various neural modules, which would be the physical analog of anigraf agents. Just as there will be advances in anigraf designs, so do we expect evolutionary advances in methods for counting votes. We follow this sequence, first with machines for plurality, then Borda*, and, lastly, Condorcet.

4.1 Plurality Voting

Let there be n alternative choices a_i with v_i of the voters preferring alternative a_i. The inputs to the n nodes in a neural network will then be the number of voters (v_i) sharing the same preference for a winner. The outcome is

$$\text{plurality\_winner} = \text{argMax}(i)\ \{v_i\} \tag{1}$$

which can be found using a winner-take-all (WTA) recurrent network whose dynamics are described elsewhere (Amari and Arbib

1977; Maas 2000; Xie et al. 2001). Note that no information about any similarity among the alternatives is captured in equation (1) because second-ranked preferences are not considered.

4.2   Borda Method

The Borda* count includes second (or higher) ranked opinions, weighting these inversely to their rank when the tally is taken (Runkel 1956; Saari 1994). The winner is then the maximum of these weighted sums. Here we use only first and second choices, with the value of the first choice doubled, whereas all second choices use their assigned weights. Furthermore, we assume that alternative choices are related by a model M_n held in common by all voters. Each voter's ranking of alternatives is now not arbitrary, but also reflects information about choice relationships. Note that the effective role of M_n is to place conditional priors on the choice domain.

Although the shared model M_n has typically been represented as a graph (G_n), it is more convenient to use the matrix M_{ij} where the entry "1" indicates the presence of the edge ij in G_n and 0 otherwise (Harary 1969).

For the graphical model of figure 4.1, we would have equation (2) as follows:

$$M_{ij} = \begin{matrix} 0 & 1 & 1 & 0 & 0 \\ 1 & 0 & 1 & 0 & 0 \\ 1 & 1 & 0 & 1 & 0 \\ 0 & 0 & 1 & 0 & 1 \\ 0 & 0 & 0 & 1 & 0 \end{matrix} \qquad (2)$$

For simplicity, we assume that the edges of M_n are undirected, meaning that if alternative a_1 is similar to alternative a_2, then a_2 is equally similar to a_1. However, directed edges require only a trivial modification to the scheme.

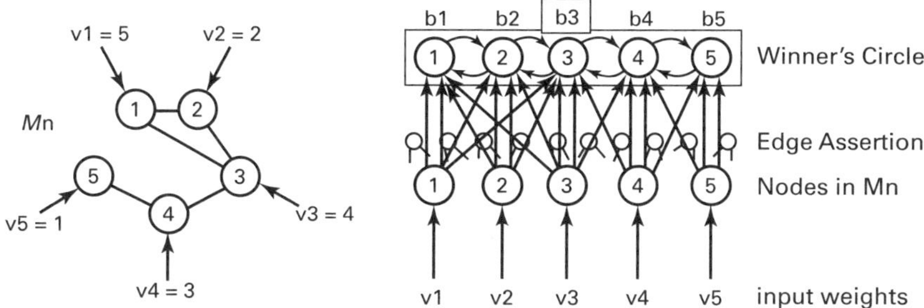

Figure 4.1

Borda* network for shared model M$_n$.

With M$_n$ expressed as the matrix M$_{ij}$ we can include second-choice opinions in a tally by defining a new voting weight v^* as in equation 3 below:

$$v^*_i = \{\, 2\, v_i + \Sigma_j\, \mathbf{M}_{ij}\, v_j \,\} \tag{3}$$

where first-choice preferences are given twice the weight of the second-ranked choices, and third or higher ranked options have zero weight. The outcome is then as follows (in equation 4):

$$\textbf{winner\_Borda*} = \textbf{argMax}(i)\ \{v^*_i\} \tag{4}$$

Figure 4.1 shows a neural network that executes this tally. This is a standard winner-take-all (WTA) network supplemented by an input layer and an edge-assertion layer that enforces the shared domain model M$_{ij}$. (The collection of WTA nodes in the "winner's circle" does not show all the recurrent connections.) The ith WTA node receives a synapse of strength 2 from the ith input node, which in turn is driven by v_i. This twice-weighted input to a WTA node is depicted by double arrows. The ith WTA node also receives synapses of strength M$_{ij}$ from the jth input. These inputs are depicted by the slimmer arrows, each of which represents the assertion of an edge in M$_n$.

In the neural network, one possible implementation of these assertions would be a layer of neurons that inhibits inputs from v_i to v_j in the winner's circle if edge $[v_i, v_j]$ is not in M_n. The highlighted WTA node 3 is the Borda* winner for the inputs v_i given in the model M_n. Note that the more common winner-take-all plurality procedure would pick node 1.

4.3   Rank Vector

A disadvantage of the Borda* Count is that a weighting on preferences is imposed depending on rank. In our simple model using only first and second choice preferences, the weightings were 2 for the first choice and 1 for the second. Let this bias be represented as the rank vector {2, 1, 0}, where the 0 is the weight applied to all preferences ranked after second choices. Then it is clear that the rank vector for the plurality method is {1, 0, 0}. But we could also invent another rank vector, {1, 1, 0}, that would weight the "top two" choices equally. More generally, a normalized rank vector will have the form {1, b, c} with $0 < b < 1$ and $c = 0$ for our simplified preference rankings. Now we see that the outcome of a Borda* procedure will depend on the choices for b, c.

4.4   Condorcet Method

To avoid specifying values for b and c, alternatives can be compared pairwise using the Condorcet procedure. Each agent then simply picks the most preferred alternative of each pair—the one with the higher rank in its preference order.

Definition: Let d_{ij} be the minimum number of edge steps between vertices i and j in M_n, where each vertex corresponds to the alternatives a_i and a_j, respectively.

Then a pairwise Condorcet score S_{ij}, between alternatives a_i and a_j, is given by (5):

$$S_{ij} = \Sigma_k\, v_k\, \mathbf{sgn}[d_{jk} - d_{ik}] \tag{5}$$

with the sign positive for the alternative a_i or a_j closer to a_k. Note that if a_i or a_j is equidistant from a_k, then sgn = 0, and the voting weight v_k does not contribute to S_{ij}.

Furthermore, as in the Borda* Count, we again impose a maximum of 2 on the value of d_{ij}, which means that third or higher ranked alternatives do not enter into the tally. A Condorcet winner is then determined by equation 6:

$$\mathbf{winner\_Condorcet} = \mathit{ForAll}_{i\,=!=j}\, S_{ij} > 0 \tag{6}$$

Although a Condorcet winner is a true majority outcome, it comes at a computational cost. For n alternatives, a complete pairwise comparison would require $_nC_2$ or $O(n^2)$ separate tallies. Thus, a neural network that calculates the Condorcet winner is superficially more complex than a network that calculates the Borda* winner.

4.5   A Condorcet Network

To reduce the computational complexity to $O(n)$, the trick is to choose a special subgraph of G_n, namely g_k, with $k \ll n$. Conceptually, the subgraph we choose is a ridge in the landscape of Borda* weights. The ridge consists of the k nodes in G_n with the highest Borda* scores. This choice is based on the observation that for connected random graphs with weights chosen from a uniform distribution, there is a 90 percent likelihood that the Condorcet winner and the Borda* winner will agree.

Specifics for the subgraph g_k
Let the Borda* rank vector be {2,1,0} as before, with the Borda* scores v^*_i for each vertex i in G_n. Without loss of

generality, label the vertices in G_n by the rank order of their Borda* score, with vertex $i = 1$ having the largest score. In cases where the Borda* scores are tied, simply choose the indexing arbitrarily among the tied vertices to create a total order. (If necessary, when ties occur for the kth and $k+1$ vertices, the size of g_k may be increased to include all vertices with scores identical to the kth ranked score.)

Definition: g_k is the spanning subgraph of G_n containing the vertices with the top k Borda* scores.

Note that other definitions are possible. For example, we could require that g_k be a connected subgraph. In this case, for some G_n, g_k may not include all the top k Borda* scores. Or, rather than ordering vertices using their Borda* scores, the top-two rank vector $\{1, 1, 0\}$ could be used. This latter choice may be more appropriate for scale-free graphs or trees with long paths between large clusters. The principal advantages of our definition are ease of computation and uniqueness.

Sketch of Network

To obtain a crude sense of the complexity of a plausible neural network that calculates a winner for g_k, we face three design challenges: finding g_k; computing the Condorcet winner for each pairwise comparison; and determining which alternative (node) beats all the others. We assume that the maximum Borda ridge (nodes in g_k) has already been found. If g_k can be unconnected as defined, then a clipping algorithm might suffice. Alternatively, if we wish to impose a connectivity constraint on g_k, then some form of a greedy algorithm beginning at vertex $i = 1$ seems appropriate. Simulations based on random graphs G_n, for $n=40$, with edge probability $1/4$ and weights chosen from a uniform distribution, show that in more than 96 percent of cases, the winners with $k = 8$ are the same regardless of whether the definition of g_k is satisfied precisely, or is found using a greedy algorithm. This equivalence might be expected, because the vertices with the highest Borda* scores will typically have the largest vertex degrees, and, hence, the greatest connectivity. In either case, this step is of complexity $O(n)$.

The second challenge is to implement $_kC_2$ pairwise comparisons (as in equation 5). The trick requires noting whether or not the vertices being compared are adjacent. Consider first the case where two vertices in g_k are not adjacent in G_n (and thus also not adjacent in g_k). Then we simply need to add the weights of the neighbors to each vertex, with the weight of the vertex itself. Then compare these two weight sums for each nonadjacent vertex to determine the pairwise winner. Note that this is equivalent to using the top-two bias vector {1, 1, 0} for each vertex, and then picking that vertex with the largest score. If the two vertices being contested are adjacent, however, note that the weight of each vertex will be added to the score of the competing vertex. Then the weights of the vertices themselves will be canceled if the top-two bias vector is used. The patch is simple: just double the weight of each vertex when adjacent vertices are being contested. This is the Borda* rank vector {2, 1, 0}. When calculating each pairwise Condorcet score, the rule is to use a top-two rank vector {1, 1, 0} when vertices are non-adjacent in $G_{n,}$, and to use the Borda* rank vector {2, 1, 0} when the vertices are adjacent. This requires making explicit whether or not the $_kC_2$ edges in g_k are adjacent or not in G_n.

The third challenge is to determine what node or vertex beats all the others. This can be handled easily by a logical AND of the WTA outputs from each pairwise comparison.

Note that although there are only k nodes in the winner's circle, in the comparator layer there will be a much larger set of roughly 2 x $_kC_2$, depending upon the tiling. This comparator

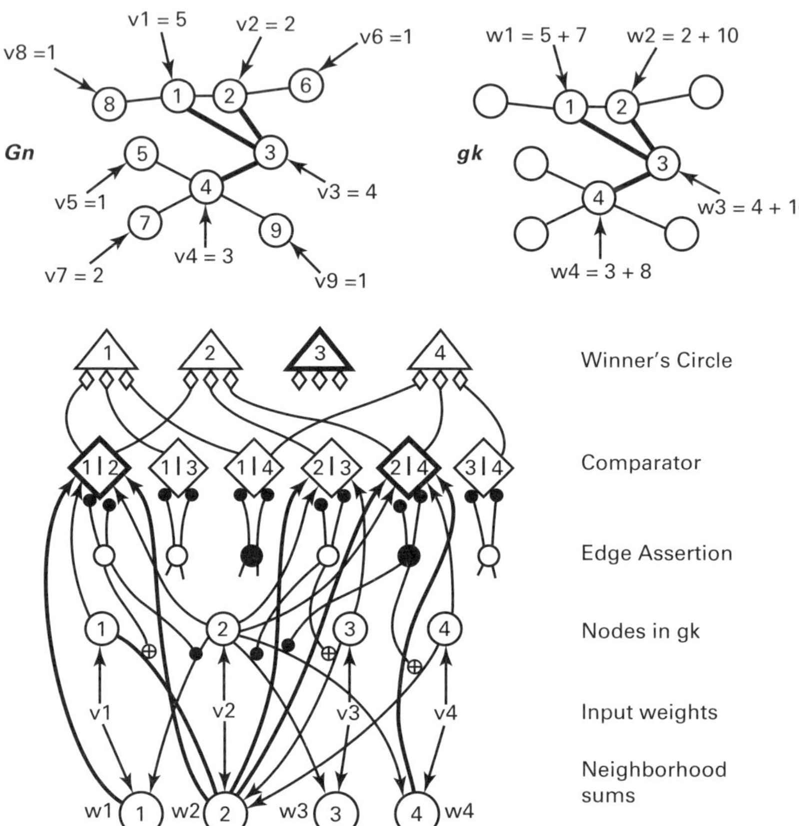

Figure 4.2

Condorcet g_k subgraph of G_n, $n = 9$, with six layers, $k = 4$ (heavy edges). Note that for Borda* counts for nodes 1, 2, 3, and 4 are respectively 17, 14, 18, and 14. Details of the computation can be found in Richards, Seung, and Pickard (2010).

layer and the comparable edge assertion layer are the critical components that govern the size of the network. If the diameter of G_n is very large, the connections required become too distant. Some hint of this problem is given in figure 4.2 for $k = 4$. This depiction also makes clear that neither G_n nor g_k appears explicitly as a graph. Rather, the connectivity is represented by the filled nodes that indicate whether the vector {2, 1, 0} or {1, 1, 0} should be applied to the paired comparison in the comparator module. This representational form has the obvious benefit that weighted edges, i.e., correlations among alternatives, can easily be incorporated by allowing analog, rather than binary, inhibition by the "edge assertion" nodes in layer 3 (small circles).

4.6   Top-cycle Tally

If there is no alternative that beats all others, then there is a top-cycle whose members, q, are those alternatives that beat all but one other alternative. Among the members of this list, identify the alternative with the largest weight (say A). Then, run a pairwise competition of A against all other $q-1$ members of q. At least one of the $q-1$ set, say B, will beat A. Hence B > A. Then, excluding A, repeat the pairwise check for B on the remaining $q-2$ members, to obtain C > B > A. Iterate through the set q to recover the top-cycle sequence for each of these "mini-tallies."

4.7   Success of g_k

Figure 4.3 shows the success rate of the k-Condorcet procedure for graphs of size $n < 250$, with different choices for k. The models M_n used were connected random graphs with edge probability 1/4. A set of weights on the nodes was chosen from

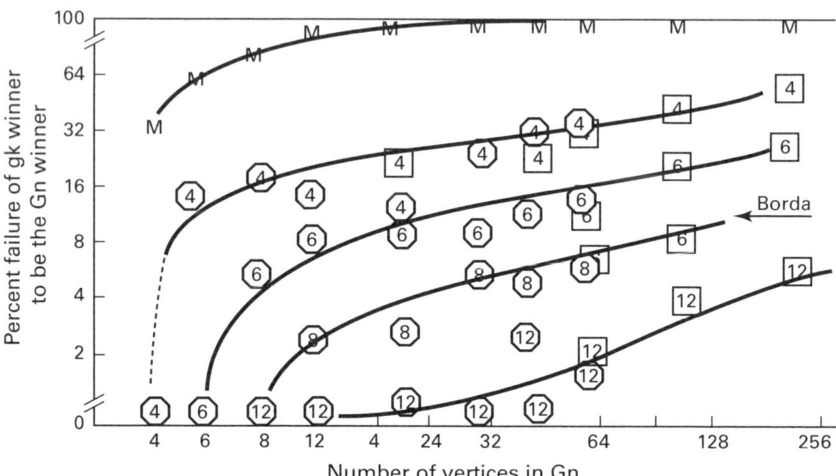

Figure 4.3

Winners for g_k compared with winners for G_n with the numbers along
the curves indicating values for k. Over the range of n, the Borda*
winner matches 90 percent of the Condorcet winners (arrow 0). The
maximum weight node in G_n is rarely the Condorcet winner for $n >
12$, as shown by the curve labeled M.

a uniform distribution. Winners were also calculated for the
same set of weights using both the plurality (i.e., node with
greatest weight) and Borda* procedures. The graph shows the
failure rate of g_k to yield the same Condorcet winner as G_n,
(There are more than one hundred trials per data point.) Also
shown is the failure of the plurality method (M) to agree with
the Condorcet choice. In contrast, regardless of n, the Borda*
and Condorcet winners differ only about 10 percent of the
time. A small fraction of this percentage is due to top-cycles in
G_n. Likewise, the principal factor for having different winners
for g_k and G_n is the presence of additional top-cycles in g_k. In
other words, when g_k picks a winner, this winner is almost certainly the G_n winner (98 + % for $k > 8$). The approximation

given by g_k is thus conservative: there are few false positives; instead no winner is chosen, unlike with the Borda* count.

4.8   The Condorcet Advantage

Both the Borda* and Condorcet networks use information about the similarity relations among alternatives by making explicit whether edges in M_n are adjacent or not. However, the Condorcet network has benefits over the somewhat simpler Borda* scheme. First, the Condorcet network finds correlations among the most significant choices when the subgraph g_k is chosen. This design thus gives the network the potential to learn priors on important correlations. Furthermore, during the learning phase, it has a clear rejection strategy: namely the presence of top-cycles. No other method mentioned previously has this kind of built-in feedback mechanism because all others always output a winner. Finally, it is not inconceivable to see the potential for such a layered network design in primitive cortical areas, or in a spinal cord, especially for occasions in which rather simple features are being aggregated.

Box 4.1
Baysian Formalizations

What is the most likely social order? A maximum likelihood calculation can provide an answer, choosing among the $n!$ possibilities for orderings on n alternatives. Following Young (1985, 1995), let $p(\mathrm{x} > \mathrm{y}) > 0.5$ be the probability that an $\mathrm{x} > \mathrm{y}$ comparison between alternatives x and y be consistent with the "true" but unobserved social order. Thus p is a measure of how often a voter's choice will agree with the underlying social order for that pairwise comparison. If there are v+ votes in favor of alternative x, and v− votes against x (i.e., for y), then the p^{v+} is the probability that the votes are chosen "correctly" and

Box 4.1
(continued)

$(1-p)^{v-}$ is the probability that the choices are "incorrect" with respect to the underlying true social order. The product of these two terms for all "x versus y" pairs in the social order is the likelihood that the alternatives x are consistent with the underlying social order. The maximum likelihood choice for the rank order is the maximum of the product, which is equivalent to choosing the rank order with the largest sum of the exponents v+.

The maximum likelihood Condorcet winner would be the top choice in the winning social order. There are two concerns with this approach. First, note the winner in the social order need not be the same as the alternative that beats all other alternatives in a pairwise comparison. Second, and more important, is the value chosen for p, which measures the fraction of a voter's choices that are consistent with the "true" social order. Prelec and Seung (2012) propose that p vary depending on a voter's expertise, as evaluated by other voters. This leads to a more robust solution for a winner. A more obvious and related problem is that voters cannot be expected to have high p values for choices deep in their preference orders. In this limit, p should go to 0.5. Hence a function like the beta distribution used by Tenenbaum (2007) is needed to enforce a value for p to 0.5 as the choices become very uncertain. The role of the similarity graph could be important here also. Together, the beta distribution and the similarity graph could act as priors on choices.

Young (1985, 1995) has also provided a treatment of a maximum likelihood calculation for the Borda procedure. Here the new obstacle is the introduction of a "rank vector," which boosts the weights on alternatives.

For n alternatives, the first choice is boosted by n. A single second choice is boosted by $n-1$. If there are k second choices all at the same level in the partial ordering of alternatives, the boost for each will be the average of the next k ranks, with the remaining alternatives apportioned in a similar manner. For any given order among the possibilities for one of the $n!$ social orders, the rank vectors obviously can differ considerably. The

Box 4.1
(continued)

rank vector is thus another variable that greatly influences the maximum likelihood Borda result. Ignoring these obstacles, for a Borda winner, we can follow the same procedure used to calculate a maximum likelihood Condorcet order, picking the first member in the most likely Borda order. Note that the Borda* procedure is less prone to the rank vector problem, because the depth of any preference order is at most three. Simulations suggest that the Borda* and Condorcet winners are likely to agree.

Part III
Cognition: Agents with Beliefs

Our earlier anigrafs might be considered as sophisticated vehicles. Although there are at least two obvious levels in their designs, namely control of an appendage that enables simple actions and a higher level choice set, their cognitive choices are rather elementary. The actions are primitive if viewed in terms of the possibilities we observe in nature. Missing are agents or mental organisms that have beliefs or models about the external world. Such beliefs form models for a reality, and require mental organisms and associated cognitive architectures that are much more abstract than those previously considered. A very compelling example of this abstraction is any inference about what lies in the "minds" of other anigrafs—specifically how other creatures view the world and what their current intentions are. With such information, the anigraf would encompass more than a physically encapsulated system with mental organisms that may attempt to control or coordinate with the actions of others. Examples include play, courtship, team formation, and alliance. Information about these intentional stances is provided in part by observed actions, but still requires recovering the belief structure of the mental models sited in the other anigraf. Our first simple example of this higher level of anigraf is a dance.

Anigraf5

Dancers: Mating Games

5.0   Two to Tango

Goals such as "fight," "flee," "evade," "approach," or "be still" have a definite cognitive flavor. Supporting these goals are the limb movements we have characterized as run, walk, turn, halt, etc. It is useful to make explicit these two levels of description. Let us attribute the observed behavior to more cognitive, mental organisms, and use lower-level agents to describe the movements of the limbs. The external observer sees only a symphony of limb movements, and from these actions infers a creature's intentions and goals. This inference is compelling (recall the labels naturally placed on Braitenberg's vehicles). For example, the direction and pace of a walk might suggest an exploration, or timidity if the direction is backward. Such movements thus provide elements for a visual language. But to understand this language, there must be a correspondence between the models of the observer and the actor. The complexity of the language increases as the actions of more parts of the body come into play, as demonstrated by head nodding or the waving of limbs. Fortunately, in the world familiar to us, a set of rather elementary movements is prevalent across many animal forms (Davis et al. 2000). They include bobbing, circling, or swaying of one's body, limbs, or

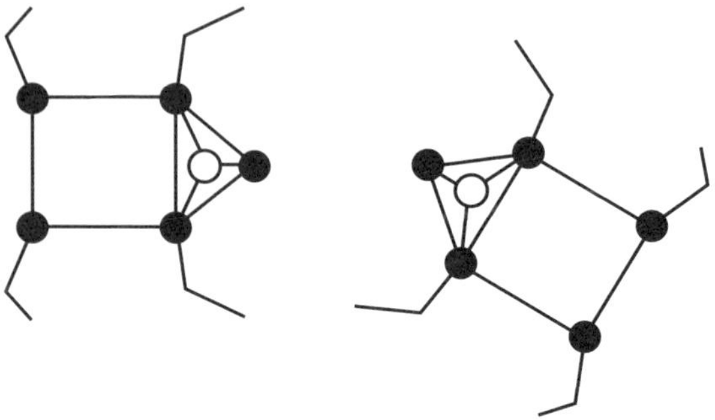

Figure 5.1

Two anigrafs engaged in coordinated behaviors. Open circle is a proxy agent. Filled circles are agents controlling limbs.

head. When two creatures of the same species coordinate such patterns of movements, we have a dance.

Definition: A dance is a sequence of movements, such as gaits, executed in coordination by two physically distinct creatures.

The challenge is to understand how the dance can be coordinated within an anigraf, where two sets of mental organisms must recognize the next (or present) goal state of one another. A simple courtship sequence provides an example.

5.1   Beyond Reflexes

It is almost trivial to build automata that execute dances. A pair of "lego robots" can perform a "cha-cha" using infrared sensors that measure distance to control the proper forward and backward movements needed (Resnick 1994). However, this solution is a simple feedback controller. At issue, therefore, is what makes our latest anigraf5 different from robotic systems, with behaviors that are no more than what one might expect from a pair of Braitenberg's vehicles. How can a social awareness common to both partners be put into place? In effect, we will need to merge two separate packages of mental organisms and their corresponding anigrafs into one integrated, social decision-making system.

To create a dance with a social awareness between the partners, a minimum of three conditions must be met:

(i) Each partner must have a repertoire of movements, or patterns of movements, in common, with similar anigraf models relating these movements.

(ii) There must be a communication channel between the corresponding agents or mental organisms that comprise the two anigrafs. By corresponding agents, we mean agents who have identical preference rankings for the various possible movements. This identity means that each corresponding mental organism understands the intentional stance of the other (but not necessarily of all other mental organisms).

(iii) When each anigraf conducts a tally, the tally must include the votes of the mental organisms (or agents) in the other anigraf. This condition insures that the decisions include both parties, and hence include an element of social awareness.

The first requirement is easily filled if each partner can execute a common set of movements. These movements then become the goal states for a set of agents, or mental organisms, which are related by two identical anigrafs. For a dance, these goal states should reflect behaviors such as flight, flight, avoid, approach, etc., and not be tied simply to a sequence of specific limb movements. For example, one goal might be "to kiss" the partner; another might be "to back away" (because of shyness or fear). For very sophisticated anigraf dancers, the choreography of such movements has the potential of telling a story (Heider and Simmel 1946).

To fulfill the second communication requirement, two agents, each associated with its own anigraf, must be linked by some kind of channel. This linkage could be visual, acoustic, tactile—whatever—as long as the channel can convey the strength of preferences of the linked agents. This stipulation is a bit more complex than the simple sensory channels found in most robots. Here, the communications must convey preferences for the actions of the system—i.e., what dance steps the two sets of mental organisms are voting upon. Agents engaged in the linkage can then pass on their assessment to higher-level mental organisms, which make the final decision.

The third requirement for an animate dance is the most crucial. The final choice of dance step must apply to both anigrafs as one integrated social system. Specifically, the tally of the preferences for actions should include all agents or organisms, combining votes from both parties. But here is the problem: there is no tally machine common to both. Hence, there is no social aggregation process. Rather, each set of agents is forced to make its own tally. At some abstract level, then, our dancers would be no different from the robots. Neither party would be performing a social activity that involved a decision made by aggregating the preference states for the entire group of mental organisms engaged in the dance. If anigrafs are to support social dancing, then each tally must include the preferences of the partner.

5.2   Proxy Votes

For a mental organism or agent in one anigraf to be a part of another's tally, the partner must acknowledge the corresponding agent's opinion as being as valid as that of its own internal agents. To solve this problem, we create a new type of agent called "a proxy." Proxies are agents who reside in one anigraf, but vote the wishes of another. Such agents will be designated using italic notation followed by an apostrophe, as in P', with subscripts added when necessary to identify the anigraf, such as P'_M or P'_W. Obviously each proxy must share a communication channel with its counterparts to enable them to read each other's preferences. So, for example, if one anigraf ,W, has reached the decision "to kiss," then this intent would be read by the "kiss" proxy agent K'_M associated with anigraf M. Such intents might be indicated by a pouting mouth or by an advanced head position (Pentland 2009). Similarly, a head lift or turn might indicate a decision to move apart or in a circle, and are read by another proxy agent T'_M. Each proxy agent K'_W or K'_M would give anigrafs the potential to know one another's intent for the next movement. To complete this potential, however, we may require that the P'-agents be linked in the same way as the action agents A in the anigraf proper. Then each set of proxies will have similar intrinsic knowledge about just what these goal states signify. Note that this constraint is no different than that imposed for our most primitive anigrafs: the edges in the graphical representation indicate that an agent "understands" goal states for the system that are different, but similar to, its own, and then includes these options when setting preference orders.

The proxy-agent relationship can have a major impact on an anigraf tally. To illustrate, consider figure 5.2, which illustrates five simple networks for one anigraf. Action agents are filled circles; the analogous proxies who report the states of a second anigraf are open circles. Here, the subscripts are omitted because we explore only the relationship between proxies $\{A', B', C' \ldots\}$ and their corresponding action agents $\{A, B, C \ldots\}$ within the same anigraf.

Consider first the two extreme cases at the right of figure 5.2, where the network design dictates whether a partner will

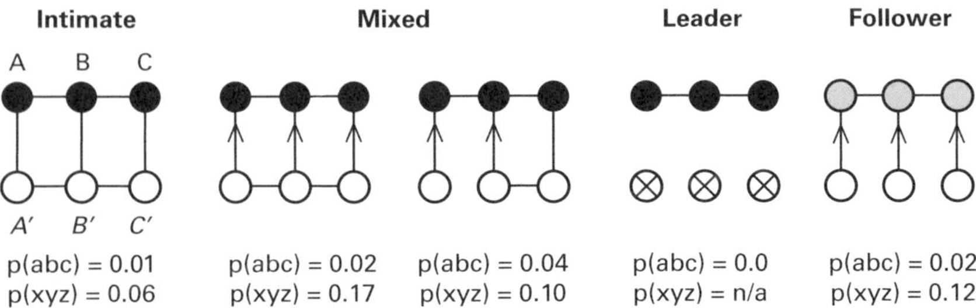

Figure 5.2

Various proxy-agent relationships for two anigrafs. Open circles are proxies; filled circle are agents. Percentages show top-cycle rates for p(xyz); those for action agents are p(ABC). Weight choices are from a random uniform distribution.

be a leader or follower. For the follower, the proxy weights will dominate the weight total for each of the action agents; for the leader, proxy weights have no influence at all. Another extreme occurs when both anigrafs share intimate, tight bonds, meeting each other halfway, with bilateral edges between proxies and agents. The depiction is illustrated at the far left of figure 5.2. Between these extremes are various combinations, two of which are shown as "mixed" networks. The probability of cycles for various proxy-agent networks provides further insight into the power of proxies using random weights on both. In figure 5.2, the probability of a top-cycle among the agents {A, B, C} is p(abc), whereas the probability of any top-cycle in the network p(xyz). Not surprising, complex networks with directional proxy to agent edges tend to have more top-cycles. When the entire network is bidirectional, as shown in the ladder graph at the left of figure 5.2, cycle probability is low. Now the state of the action agents {A, B, C} contributes directly to the proxy votes {*A*', *B*', *C*'} when a tally is taken. However, there is a curious twist: if an anigraf's own beliefs influence how another anigraf's intentions are "read," then

Box 5.1

> Proxy agents are obvious analogs of mirror neurons, first reported by Rizzolati in 1990. Although these early findings were controversial, there now appears to be general agreement that mirror neurons serve an important function in establishing empathetic relationships. Ramachandran (2007) also suggests that they may provide a basis for self-awareness. Note however, that proxy agents in anigrafs have a very specific function: namely, to mimic external states so they can be evaluated by a creature's tally machine.

non-cooperative behaviors may result even for otherwise stable networks. This seems inappropriate. Therefore, regardless of the fact that top-cycles may increase, the more plausible proxy-agent networks should have directional edges so the strength of proxies will influence the weights of action agents, but not vice versa.

Regardless of the stability and instability of outcomes when proxies participate in a tally, these new agents add the ingredient of a social awareness held in common between two partner anigrafs. In the ideal case, when the partners are twins and have the same anigraf model, with perfect proxy readings, then there is a strong "meeting of the minds." Each anigraf has been augmented in an equivalent manner such that when the Condorcet tally is conducted *in parallel, within each anigraf*, then the outcomes will reflect the intentions of both anigrafs. A shared social consciousness emerges.

5.3   A Duet

Figure 5.3 shows a more detailed abstraction of two anigrafs that might participate in a duet. The solid circles represent the internal agents who effect actions, and the open circles are the

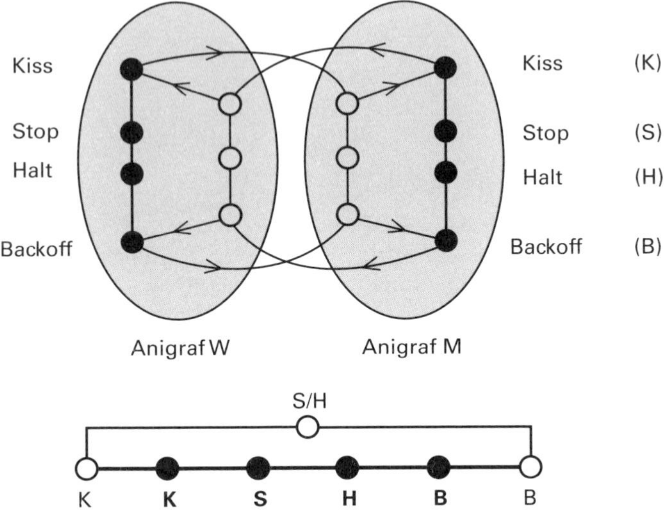

Figure 5.3

Model underlying a "kissing" dance.

proxies. The actions are to move forward "to kiss" (K), to stop kissing (S), to move apart (B), and to halt the backward movement (H). Figure 5.3 illustrates the relationships among these actions. The figure depicts the underlying "mental model," which is equivalent to a four-chain, plus the associated proxies. Cyclic outcomes supporting a coordinated dance are now easy to arrange.

Our kissing dance example is basically a cyclic "cha-cha" with four action states: K, S, H, and B. With directed proxy input, a three-cycle will occur by chance about 20 percent of the time, and the rarer four-cycle occurs for about 5 percent of random weight choices (as in figure 5.4). Whether or not all or none of the proxy agents are connected to each other makes little difference. These percentages are high enough to allow a simple dance to be learned, even with trial and error searching for appropriate weights on nodes.

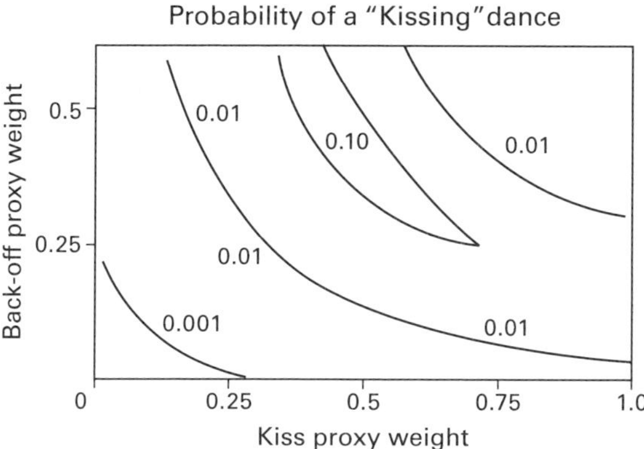

Figure 5.4

Probability of a kissing dance for different proxy weights.

One might now ask again how this "cha-cha" differs from one enacted by lego robots or vehicles? First, the complete "cha-cha" requires a K, S, B, H four-cycle in a graph that need be only partially directed. Because there is bilateral communication between action agents, with shared information included in each tally, the outcome is a social process. Although the proxies might be regarded as robotic-like sensor inputs, their effect is quite complex and not modeled by classical controllers. The state transitions arising from the pairwise tally are difficult to predict. (See figure 5.5 for a relevant phase plot.) Outcomes are not simple additions of an agent's voting power and the proxy weights. For example, our paired anigrafs *could* engage in a cyclic cha-cha even if one of the agents K or B, but not both, had zero voting power! Alternately, if both votes for K and B are simultaneously near their maximum weight, then there will be no dance cycle. This makes intuitive sense: mental organisms cannot collectively decide to engage in kissing and breaking up simultaneously. The cyclic dance, then, succeeds only with some measured

Figure 5.5

Phase plot for a kissing dance of four-steps (equivalent to a directed chain with two directed proxy agents at end of a four-chain.) Textured areas are top-cycles. Weights are (4, x = 0,10; 5, y = 0–10, 0, 6). Note regions of four-cycles.

input by both sets of proxies. Anigraf social behavior is quite nonlinear.

5.4   Social Twins

Most creatures use many different channels and modalities to communicate. Bees, for example, have at least three different channels to indicate food location, quality, and quantity: body motions (two different patterns that include body contact), wing beats, and sounds. Ants have an array of chemical phero-mones (Wilson 1971). The number of different signs used by primates is almost countless (Darwin 1859; Bond 2000). Add-ing more specialized communication channels allows the

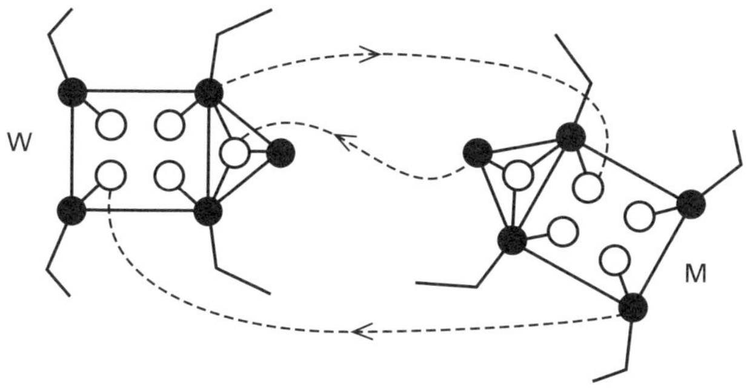

Figure 5.6

Two anigrafs engaged in a dance. Open circles are proxies; filled circles are action agents.

Box 5.2
Hypnotic Dreamers: a byproduct

Two fascinating situations arise when proxy agents come to dominate an anigraf's behavior. In both cases, the proxy communication channels would remain active with strong inputs, but the action-agents themselves would be very weak. In such a state, the anigraf has become subservient to the proxy demands. If a second anigraf's desires now dominate the proxy channel inputs, then that anigraf is completely controlling the behavior of the other. In a second version, we could also imagine random external inputs controlling the proxy values, but with no second, dominant anigraf present. The situation is now like a dream, whereas the first resembles a hypnotic state.

Could there be an evolutionary advantage to losing some control over one's active agents? For example, what if the current tallies lead to chaotic, maladaptive behaviors? Or what if there is some kind of frustrated decision-making, incompatible with the physical world? Relinquishing control by active internal agents might then be beneficial. Viewed in this light, proxies could be used to produce a calming effect on a creature's behaviors. Examples of such beneficial proxy control might include grooming, stroking one's skin, rocking motions, soothing music, etc. Under the right circumstances, suitably tailored proxy agents could easily dominate an anigraf's tally. Could dreams and hypnotic states have a set of agent-types in common?

coding to be greatly simplified, in the limit reducing to sending a single yes/no bit. But this takes place at the expense of adding more proxy agents. The great benefit, however, is that the aggregation process becomes more social: each anigraf now has greater participation in the partner's tally. Schematically, two interacting anigrafs of the same species might be depicted as in figure 5.6. Here, for clarity, only three proxy communication channels are indicated by dashed lines. (Line segments without nodes are included to indicate limbs controlled by lower-tier agents.) Like all other anigrafs, M and W each have their own separate tally machine. But each tally will yield the appropriate social outcome for the actions of the pair. At least this will be true when M and W are identical twins, initialized with the same weights on identical action agents. However, even with slight differences in anigraf forms and proxy reliability, anigrafs now have a sense of understanding the intent of others, allowing them to engage in a variety of group activities. These are first steps toward social communication.

Anigraf6: Planners
Event Sequencing

6.1　A New Level of Complexity

A dance requires choreography. For humans, we create a variety of dance sequences from a rather limited set of limb movements. However, for many fish, birds, and other lower animal forms, there are "innate releasing mechanisms" with the choreography "built in" (Lorentz 1982; Tinbergen 1951). This is especially clear in mating behaviors, where consummation requires first various displays that arouse interest, then contact, and finally intercourse. Often the environment plays a key role in choosing the correct action where one act sets up the preconditions for the next. Whether the behavior is the building of a spider's web, the burying of a nut, or the moving of a boulder, it entails steps where one act changes the external state that in turn allows the next act to proceed. The situation is much like when a cook prepares a dish: the ingredients are laid out on the counter, and once a sequence of steps is initiated, certain operations become obvious, whereas others are clearly not (O'Regan 1992).

If anigrafs organize subgraphs into cliques, or develop new cliques, not only may we have pandemonium in that many sets of mental organisms would be clamoring for attention, but we might also have multiple, conflicting demands on the same effector system. Part of the problem arises in that each clique is conducting its own tally at its own rate. This prohibits any coordination among the active mental organisms. We need machinery that adjudicates sequences such as those in the kissing dance. If the main task is to "get food," for example, then there is an obvious sequence: look, approach, grasp, eat, etc. If there is an activity that requires coordination between two anigrafs, again there will be a proper sequence of steps that must take place to consummate the goal. We need agents and mental organisms that set up these sequences. These mental organisms will be called "brokers." Note that brokers control the category of an action, but not the action itself.

6.2   Depiction

Figure 6.1 shows a possible setup. There are three cliques of action agents {Ai}, {Bi}, {Ci}, each with their own tally machines Ta, Tb, Tc. These cliques differ in the kinds of actions they initiate. Clique A, for example, may govern gait type (walk forward, retreat, etc.); clique B may determine how things are grasped (pickup, release, push); and clique C might govern whether something should be eaten, or simply bit into, or just poked at. Note that for a given set of categories, different sequences are possible. If the goal is to eat an object, the clique sequence would be A $\rightarrow$ B $\rightarrow$ C, whereas if food is offered by another anigraf, the sequence might be B $\rightarrow$ C $\rightarrow$ A. The role of the brokering agents is to decide which sequence of cliques is appropriate. Thus, each broker has a model for the relation between a set of events. At the top of figure 6.1 are three ellipses that indicate the mental models of three brokers. Within these ellipses are events (indicated by starred diamonds) whose relationships represent the mental model held by the broker. Each broker will conduct its own tally to choose from the events within its set. There is also a global tally T^* conducted among

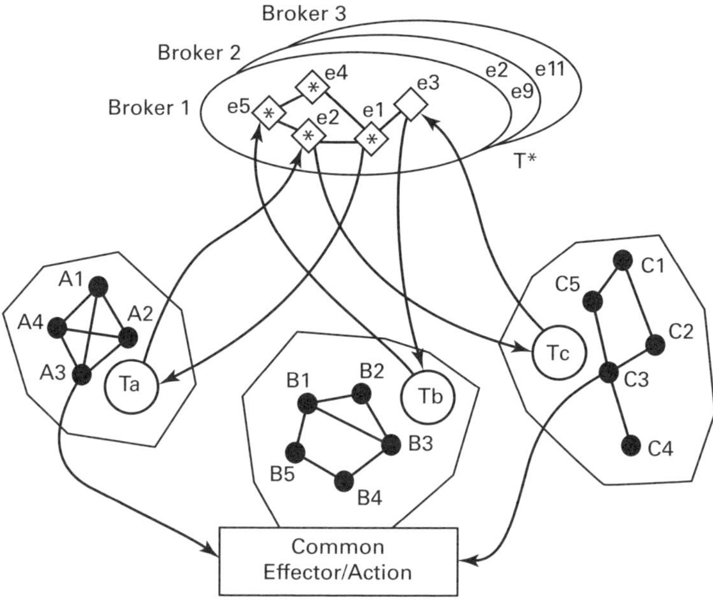

Figure 6.1

Three brokers controlling three different anigraf action cliques. Note that each level has its own tally machine (T_i or T^*).

all the brokers to choose the broker who will control the sequence of action agents. The weights for this global tally are set by desired goal states. Not shown in Figure 6.1 are the graphical relations among the goals of the various brokers. These relations may be dictated in part when brokers share the same event, or when their event graph structures are similar.

6.3 Event Sequence Control

Once the tally T^* is taken and a broker is chosen, how is the particular event sequence determined? Several options are possible.

6.3.1 Choreographed Cycles

The most obvious solution to event sequence control is for each broker to have one pre-programmed sequence that runs when initiated. For example, as illustrated at the lower left of figure 6.2, a sequence of A2 → C3 → E1 → D5 might ensue regardless of weights on the action cliques. However, to give our brokers more flexibility, we could have the sequence set by a social cycle among the possible events. In this case, different sets of weights would generate different event sequences. This control structure is simply analogous to an open loop "dance," or, at a lower level in the anigraf hierarchy, to a sequence of programmed limb movements. If brokers had ring graphs with an odd number of nodes (events), such as in a pentagon, then any triple of event sequences is possible given the proper set of weights. For small rings, choosing between different cycles becomes trivial. For example, for the bidirectional pentagon in figure 6.2, the only difference between cycles ACD and BDE is that the weight set has been shifted clockwise by one vertex.

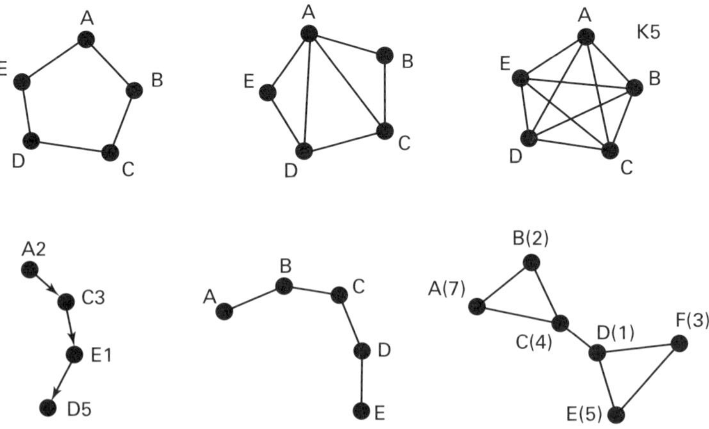

Figure 6.2

Six different cliques that can generate different action sequences (i.e., top-cycles).

6.3.2 Inverted Condorcet

The obvious disadvantage of cyclic sequences based on a fixed set of weights is that the program will run to completion regardless of whether the actions are appropriate. A more desirable option would be to allow the weights on the events (or cliques) to play a role in choosing (or verifying) the sequence. This is easily accomplished by using an inverted Condorcet tally. First, initiate the event that is beaten by all others (i.e., the biggest loser). Next, since this alternative has been satisfied, set the weight on this first event to zero. Then, pick the next loser, etc., progressing through all events under the control of the broker. We illustrate using a directed chain graph (the lower left of figure 6.2). If the chain is not directed, however, many variations on sequences can occur even with this very simple procedure. Using the same weights as before, the sequence for the undirected chain [A ... E] will now be A $\rightarrow$ C $\rightarrow$ D $\rightarrow$ B. The difference results from the nature of the Condorcet tally, which considers contributions from neighboring nodes.

6.3.3 Strictly by Weight

If we wish to have only the weights on events control the tally, then contributions from neighbors can be thwarted if the broker's graphical relationship between events is the complete graph K_n. (K_5 is illustrated in the upper right of figure 6.2.) Because all nodes are connected to each other, in any pairwise comparison between two nodes, all remaining nodes are neighbors to the combatants. Thus, they will be indifferent in their preference, and the node with maximum weight will win. If the weight on this node is set to zero, the node with the next largest weight will win, etc.

A variant of this procedure would be when only one node covers all others, as shown in the top-middle of figure 6.2. Event category A now wins the first vote roughly 70 percent of the time because it receives support from all other nodes. But now what if the weight on A is zero? The A node will still win 16 percent of the time. Thus, an ordered sequence of weights is not guaranteed to produce a comparable sequence of node choices.

6.3.4   Constrained Sequences

The covered graph is only one of many examples where the graphical form of the relationships among events held by a broker is the important constraint on event sequences. Consider the bowtie shown at the lower right of figure 6.2. Let the weights on nodes (A, B, . . . F) be respectively (7, 2, 4, 1, 6, 3). When a tally is taken between any element of the triangle ABC versus any element of the triangle DEF, the winner will be that triangle having the greatest weight sum (or lowest, if inverted Condorcet is used). Thus, the first winner is A, which beats all its neighbors, as well as D, E, and F which together can muster only $1 + 3 + 5 = 9$ votes against A's total of 13. But once clique A is triggered, and its weight drops to zero, then the voting power will pass to the right triangle DEF. Clique E becomes the new winner. Continuing this process for the weights shown will elicit the sequence $A \rightarrow E \rightarrow C \rightarrow F \rightarrow B \rightarrow D$. This particular graph will favor alterations between the two triangles, which might correspond, say, to a sequence of various types of symmetric body motions, each followed by different limb motions.

6.4   Beyond Brokers

The development of hierarchical anigrafs, with brokers controlling event sequences, has now introduced a level of complexity that presents a major hurdle to further lucid developments. Additional proposals for embodied anigraf

Box 6.1
Spanning the Brain

> The introduction of event brokers has consequences for brain design. The scheme clearly requires reciprocal long-distance connections between the separate tally machines of various action cliques and the site where the brokers reside. Yet to facilitate network building, brokers must be in proximity to each other, conducting their own tally as to the winning broker. For example, there is correlation between an edible object, what it looks like, its taste, and the creature's ability to walk to it and grasp that object. In such a scenario, vision, taste, locomotion, and grasping must all come together in concert, even though cliques that control these actions may be located in quite different places. If we have m action cliques of agents, then two-event pairing of cliques will require $_mC_2$ brokers and edges (to the cliques). However, this is a potentially formidable number. Yet, if at least some connections were local (to the brokers), then the control becomes more manageable because only neighborhood communications between agents executing similar sequences of actions would be needed. Our creature would then need only m bundles that span its brain going from the brokering area to the cliques that carry out separate tallies for specific actions.

designs can only make the construction more opaque, endangering a simple picture of creatures whose minds are social networks. Nevertheless, there is one more insight that can be gleaned from our brokers. What if the set is not embodied in one physical entity, but rather is seen as a collection of different peoples, each with their own game plan. Yet together, they would form a society in the popular sense. What kinds of behaviors might be expected?

Anigraf7
Explorers: New Worlds

7.0 Expanding Horizons

As anigrafs become more engaged with their worlds, new relationships among events will be discovered. Some of these will involve inanimate objects and actions; others may follow from interactions with new species of creatures, perhaps with different embodiments, and hence with different internal models. These new relationships will require new, or revised, models that are appropriate for the context. Two options are obvious: edges can be added or deleted; or, nodes can be created or destroyed. Both come at a cost. If edges are revised, then the altered anigraf, appropriate for the new context, may not have the proper form for the former context. Similarly, when nodes are added or deleted, what happens to the discarded mental states? Also, how are new mental organisms or agents created? And how are preference rankings affected? For example, the state associated with the addition of a new node becomes a second-choice preference for all the original nodes to which it is connected. When the degree of connectivity is high, this addition of a single node can cause a major change in the anigraf network, and hence its model for actions. Therefore, nodes and edges cannot simply be added haphazardly.

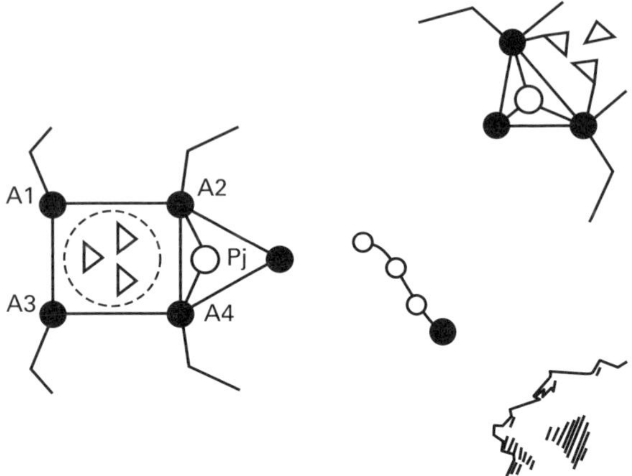

Figure 7.1

Creatures in anigraf world. Ai are action agents; Pj are proxies.

In order to avoid disrupting actions of successful anigrafs, one simple solution would be to duplicate the original anigraf, and then revise relationships properly for any new context. If certain anigraf forms were very common (e.g., rings, chains, trees, etc.), it is easy to imagine a reservoir of these forms available for use. To facilitate the development of new models, perhaps each active anigraf model should have associated with it a stack of similar models with unassigned nodes. Then a context revision would simply require a shift to another level in the stack, with an analogical labeling of these new nodes.

Whether current models are revised by the addition of a few nodes, or by labeling analogous forms, the anigraf's graphical structure rapidly becomes larger and more complex. Certainly on occasion there may be deletions of edges or nodes, if relationships are formed in error. However, more likely, nodes will be deleted only when one goal state is refined to offer several alternative subcategories that have become desirable.

Furthermore, whenever sensory inputs or motor outputs are augmented, we can expect new agents or mental organisms to be added to represent these states. Changes in an anigraf's form are thus almost exclusively effected through node additions, rather than deletions.

7.1   Free Agents

To add nodes, the anigraf must possess some "free agents." These are agents whose preferences are initially unassigned, but are capable of acquiring a new goal state and learning that state's relation to present goals. Free agents initially have zero voting power. They are depicted as triangles. In some cases, several free agents may be grouped together to form a clique, already having tentative (unspecified) relationships to one another.

One might assume that if "free agents" have zero clout and no preferences, then they can be connected arbitrarily to action agents without changing behaviors. However, this is not true, as we know already from our examination of proxy-agent relations. Rather, the development of more complex anigraf models will be constrained both by the model's current form, as well as how free agents are added to the network. Consider a single free agent that is bidirectionally connected to a large number of agents. In any Condorcet tally, this free agent with no goal state could have a very large weight score resulting from the addition of all its neighbor's weights, and this total could exceed that of any other agent. At the other extreme, adding free agents with bilateral connections to action agents can also increase the probability of no consensus (see figure 5.1). Not only will the probability of top-cycles increase, but even more drastic, the chance of a free agent "winning" the Condorcet vote becomes very high as the free agent population

Table 7.1
The strong adverse effects of free agents on winners.

Percent of Free Agent Winners						
#Free Agents	0	2	4	8	16	32
Random G_6	0	4	9	15	20	25
Chain C_6	0	18	28	40	55	70

increases. Therefore, if one desires that anigrafs have the potential to acquire new models for different contexts, constraints are imposed by both the size of the population of unlabeled agents with respect to those already labeled, and the form of connectivity between inactive, unlabeled agents and active agents.

Consider a set of six action agents, arbitrarily linked to form a random anigraf G_6. Now add at random to any of these action agents an increasing number of free agents. As shown in the second row of table 7.1, if we add sixteen free agents with no voting power, 20 percent of the tallies will be won by one of the added sixteen free agents (although they have cast no votes). If the fixed set of action agents has the form of a chain C_6, then the percent of free agent winners becomes disastrous even when only four such agents have been added with no voting powers.

A possible remedy is to require directed edges between free agents and action agents. Then any Condorcet winner will almost certainly be one of the action agents. Consensus is still not guaranteed, however, because top-cycles can still emerge. As shown in figure 7.3 for chain graphs with six action agents, as a small number of free agents are added with directional edges to action agents, the no-winner probabilities will increase, but then decline to about 10 percent near the level of directional chains. Clearly the choice of model is a very significant factor. Covered graphs, such as a "wheel" of six action agents,

Figure 7.2

Phase plot showing winners for an undirected chain of six agents, with four free agents of zero weight randomly connected to the six agents. Textured areas show top-cycles. Some cycles include free agents. Weights are (6, 1, x = 0–10, 2, y = 0–10, 4, 0, 0, 0, 0).

will have guaranteed winners with no top-cycles; however, a six-chain can double the rate of no-winners over random graphs. For the neuroscientist, these consequences may be disturbing, since two neural networks that on first appearance appear identical could actually have quite different behaviors.

7.2   Imposing Constraints

Of particular interest is the ideal number of free agents for any given number of action agents. Three constraints seem obvious: First, every action agent should have associated with it at least one free agent, thus giving this agent some potential for expressing a new relationship. Second, any one free agent

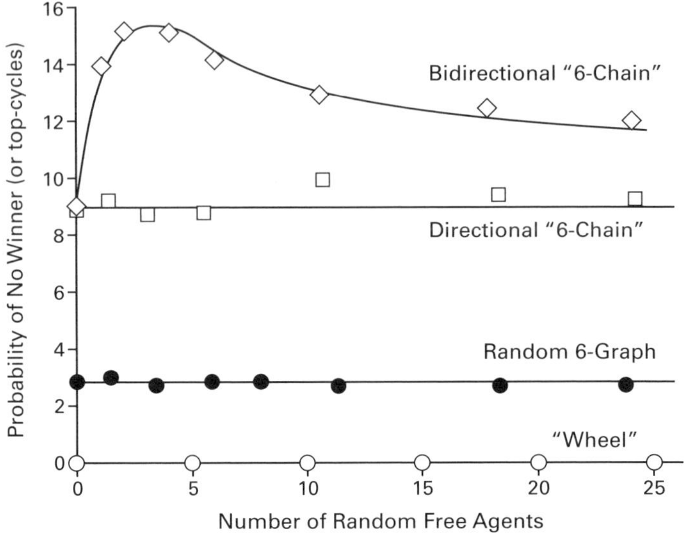

Figure 7.3

Illustrates how increasing the percent of free agents will affect the no-winner probability.

should not be associated with a large number of action agents, or any specificity in a new acquisition is diluted. Third, top-cycles should be minimized or lie within some reasonable bounds (say 10 percent). Figure 7.3 suggests a possible satisficing solution of about twice the number of free agents for any set of action agents. For the case of six action agents with similarity relationships chosen at random ($p = 1/2$), top-cycles are limited to about 3 percent maximum. For this set, we need at least six free agents. Adding twelve gives good odds ($p = \sim 3/4$) that each agent will have at least one free agent associate, if pairings are chosen at random. Thus, for random graphs, there should be roughly twice as many free agents available as there are action agents. For special graphs, such as chains, the ratio may be higher; and for highly covered graphs (not shown), the ideal ratio may be lower. Such constraints remain unexplored. Nevertheless, we can visualize a reticular net of free agents surrounding any set of action agents. Most of these free agents will presumably begin their life as inactive appendages to active agents. But others may be lurking nearby, with links to their cousins that will help speed up any model development process.

7.3   Graphical Evolution

As previously explained, new goals and action states should not totally disrupt mental models that have had past success. Rather, new goals should augment or correct flaws in present models. This constraint implies two guidelines: (i) that new goal states (together with their proxies) should be added sparsely, with few initial edges; and (ii) that revised relationships between present action agents should also lead to minimal changes in outcomes.

7.3.1   Node Additions

Referring to the upper-left panel of figure 7.4, let the anigraf form, or model M_n, be a simple chain C_n—i.e., the agents or mental organisms have a set of relationships that lie along a path. Then adding a free agent (u) to one of the active agents will introduce a minimum of cycles, as compared with an "en masse" augmentation shown at the far right, where now many additional rings appear as subgraphs. The next minimal increment would be to add another free agent (u'), perhaps to the same node (A) or elsewhere. However, adding trees to active agents is also plausible (e.g., "w"). Our simple linear chain rapidly expands into a much larger tree. Along the way, the anigraf may recognize new similarity relationships between action agents, creating new edges such as (u, v), which may share a

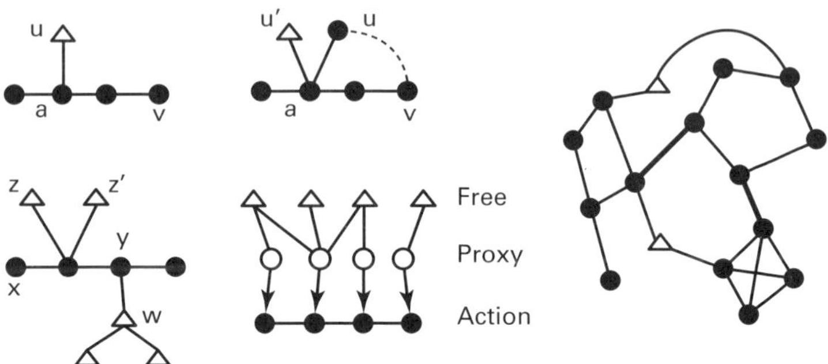

Figure 7.4

Augmenting tree forms with free agents.

new feature dimension. Behavioral cycles in outcomes now become a larger threat. Also complete subgraphs K_n may be formed, such as if x is linked to y to create K_3, or the K_4 subgraph shown at the right of figure 7.4. Because such subgraphs add little differentiation among the included agents, K_n subgraphs for $n > 3$ will be replaced by a new node. Such informal rules for augmenting any anigraf can be implemented rather easily. One method is to have a reticular net of free agents, as previously described. However, another, perhaps more attractive, method would be to convert proxies to action agents that already have similar goals. (See the lower-middle panel of figure 7.4.) This alternate has the advantage that in the unmodified network, free agents (with zero voting power) can be connected arbitrarily to proxies without changing outcomes. Furthermore, since proxies typically are linked (directionally) to only one action agent, we satisfy the desired three minimum conditions mentioned at the outset. Once a proxy is reassigned as an action agent, its connectivity to its previous host is changed to bidirectional, establishing the desired new similarity relationships, and any previous proxy-proxy relationships would wither.

7.3.2 Edge Additions

In any physical embodiment, distance between nodes in a network becomes a factor in the ease of establishing connectivity. If a new belief or goal state is added, what insures that this state can indeed be linked to present agents or mental organisms that share similar preferences? If the similarity in goal states is derived from a similarity in the interface device (sensory or motor), then one might expect some physical proximity between the now newly specified agent and its most similar existing action agent. Although the "proxy" conversion satisfies this condition, other desired linkages may not. This "physical proximity" constraint implies that any sensory interface or effector connections of free agents must already be biased to parallel those of existing action agents. Again, proxy conversion is a straightforward way to support this design constraint—if free agents are available to serve as subcategories of preferences of present goal sets. For example, the baby anigraf may not distinguish between types of locomotion, such as crawl, walk, run, etc. Free agents "in the neighborhood" of the "approach" proxy could be added to make such distinctions in the kinds of approaches (or withdrawals). This design implies that the learning of new behaviors is critically dependent on behaviors already in place; see Marler (1987) on differences in learning capabilities of pigeons and rats. The result is a very stable augmentation in behavior, with a morphology of the mental model that favors a tree network. Such an anigraf form has a very high likelihood of unique winners.

7.3.3 Giant Components

As we continue to refine the beliefs and goals of agents and mental organisms, any small clique of action agents rapidly grows. This growth raises the possibility that many more relationships will be found between members of different cliques.

Certain loud sounds may be associated with very bright (lightning) flashes, or temperatures with certain visual colors. Powerful running requires more than just leg motion, but also hip movement and foot exertion for added lift. These in turn affect balance. So the anigraf networks can easily explode well beyond a simple tree, more likely resembling some large random graph. To find food, we must use vision, taste, grasping, locomotion, etc.—a complex, context-sensitive network of quite different agents. Three very serious problems emerge:

(i) One tally machine for the millions of agents potentially involved in decision-making seems very unlikely: the cycle time involved (for neurons) quickly becomes prohibitive.

(ii) Large networks created by combining cliques would be expected to have increasingly large distances between nodes (and hence agents). Therefore, if agents are to share preferences: the distance of their communications must increase well beyond nearest neighbors; and the set of preference options for any one agent must increase as well, otherwise communications will be meaningless.

(iii) Large "random" graphs with "noisy" agents are more likely to lead to cyclic outcomes with no clear winner.

We can overcome part of the second objection of network size by placing constraints on the form of very large graphs, and thus the way cliques and free agents are added or merged. If graphs evolve into random graphs, where edge connections have a fixed probability, then for very large graphs, the distance between any two agents will be at most two edge steps. A related, but not quite as useful, result can be obtained by ensuring that the graph is a small world graph, which has a fractal distribution of distances between any two nodes (Watts and Strogatz 1997). See also Rossi et al. (2011). These "solutions," however, lead us directly into the third objection: noisy agents can be devastating in reaching decisive outcomes. Furthermore, there is still no limit on the number of preference options any one agent must be capable of sharing. And, finally, this solution does not solve the problem of tally time. Somehow, cliques of agents must remain small—probably on the order of less than a dozen or two. This would mean that tallies would be conducted simultaneously with many cliques shouting for their turn at controlling behaviors.

7.4   Anigraf Fitness and Information

The graph of the similarity relations among alternative choices is a key component of an anigraf. This similarity graph captures information about the context of the decision space. The more information available, the better the decision outcome reached by the daemons will be. Therefore, the fitness of the anigraf will be optimized. This information will reside in the pattern of edges in the graph—not in the vertices of the graph. The vertices express first-choice preferences of each daemon and are typically fixed; in contrast, the edges reveal relations about the structure of the context and may change. Thus, unlike most studies in population and evolutionary dynamics, which modify vertex types, anigraf fitness is expressed by the pattern of edges in the similarity graph. An exception would be if the daemon were a "free agent," with initially unassigned preferences.

To make the jump from the evolution of vertex types to the evolution of edge types, we can recast the anigraf G_n into its line graph $L(G_n)$.

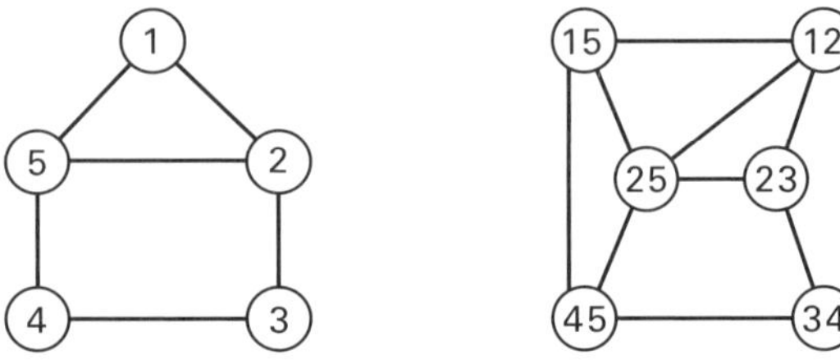

Figure 7.5

The graph G_n (left) and its line graph $L(G_n)$ (right).

Definition: The vertices of the line graph $L(G_n)$ for a graph G_n represent an edge of G_n, with two vertices of $L(G_n)$ being adjacent if and only if their corresponding edges share a common endpoint.

With this transformation, many studies of population dynamics on vertex types can be recast in terms of studies on edge types. The elegant study by Lieberman, Heuert, and Nowak (2004) would be one example. The burden then is to define a measure of fitness relevant to anigraf goals. For example, if we desire to invoke cyclic behavior, then we would seek a graphical form that maximizes the probability of top-cycles. Alternatively, if a quick decision or judgment is needed, finding the winning daemon might be the objective—the opposite of a top-cycle search. A simple solution to the quick decision problem would be a complete graph for the alternatives. This trivial solution may be maladaptive, however, because all alternatives would be equally similar, and the outcome would be the daemon with the maximum weight. Clearly there is no (or little) information about the domain. More productive would be to find the graphical form carrying the most information about the context.

The first attempt to measure the Shannon information content of a graph was Rashevsky (1955), who examined the topology of the graph. Since then, many proposals have been made, largely in chemistry, biology, and sociology, with the most recent proposals in network physics. (For a review, see Mowshowitz and Dehmer 2012.) These measures range from Kolmogorov complexity to probabilistic, the latter being the most relevant to anigrafs. Ideally, any proposal should include some sense of the probability of graphical patterns in the domain of interest. However, such patterns are not directly available to an anigraf. We clearly have entered a new area of study that has much room of exploration and theory.

Anigraf8
Alliances: Coordinating Diversity

8.0   Heterogeneity

Consider a society of identical anigrafs. Then, unlike the connectivity shown in figure 8.0, the social map will be the complete graph. However, with such homogeneity, there will be accurate, intelligible communications among members of the society. Sacrificed will be the ability of the group to see the world from different vantage points. Behaviors will be very predictable. Clearly, the flexibility and adaptability of a society depends upon its members possessing a range of talents to execute a host of different tasks (Page 2007).

Fortunately, given even minimal environmental pressures, offspring are not carbon copies of their parents. As the number of models increases to deal with new events and challenges in the world, an offspring's view continues to depart more and more from the parental path. Diversity evolves, even in the presence of a common core. Examples include insect societies, football teams, or a jazz combo. Yet in the presence of diversity there is still a unity among members through an almost unconscious bond. How different may any two members be, and still experience such a unity? The answer sets the stage for examining how a population of anigrafs with different models may be able to form productive alliances.

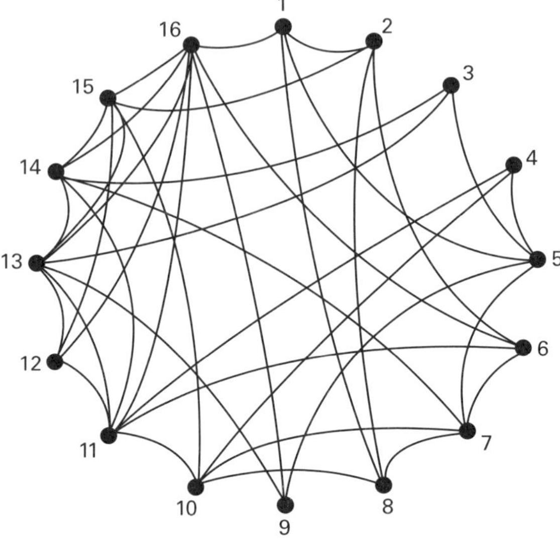

Figure 8.1

A social network showing the connectivity among sixteen different anigrafs. The links are sparse because the anigraf similarities are presumed to be weak.

8.1   Similarity and Dissimilarity

In the world of anigrafs, similarity and dissimilarity are tied to graphical forms, or more precisely, to the derived preference orderings. These preference orderings reflect goals and choices related by the mental organisms that comprise anigrafs. If two anigrafs have mental organisms with the same set of goals, then these anigrafs will be identical only if their graphical forms are the same. An opposite extreme would be two anigrafs with the same goals and choices, but with preference orderings derived from a very different model. To illustrate, figure 8.2 shows two graphical models that are complementary: each are subgraphs of K_4, and together form the complete

Figure 8.2

Two complementary graphs and their preference orders.

graph. None of the preference orderings agree. Thus, even given the same set of weights on alternatives, the Condorcet choices typically would differ. Our aim is to have diversity in anigraf forms, yet still have enough similarity among members of the group so that these differences do not prohibit coordinated actions. This is the essence of building an integrated team of players.

To capture our intuition about similarity and diversity (or equivalently, dissimilarity), we can measure the fraction of agents or mental organisms in two anigrafs who have identical preference orders. This measure automatically takes into account dissimilarities introduced by differences in goals.

Definition: The similarity, $S(Kd)$, between two anigrafs having n and m vertices is the fraction of all $(n+m)$ vertices that have the same preference orderings, where "Kd" is the depth of the preference orderings.

The patterns in figure 8.3 exhibit two pairs of anigrafs with similarities of 1/3, with preference orders based on neighbors only—i.e., Kd = 2. (A, A′, P, P′, etc., are identical goals in the anigrafs being compared.) To generate these anigrafs, the edge set of the left member was chosen with probability 1/2. A core for the second anigraf was then created by freezing vertices A, B, and P, Q, keeping their same edge sets. The remaining four

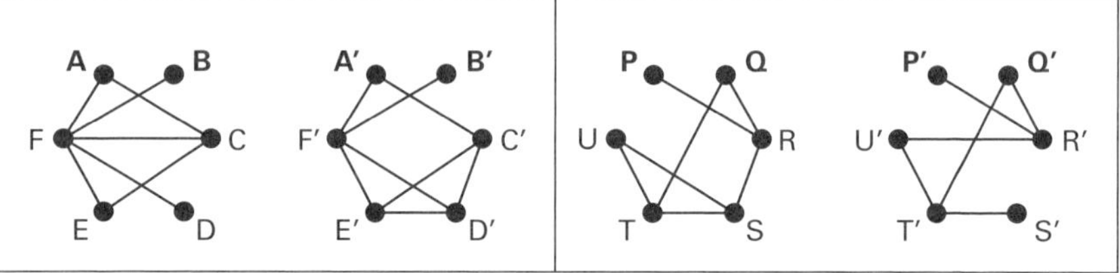

Figure 8.3

Two examples of one-third overlap with Kd = 2.

vertices were then assigned edges with probability 1/2 using the reduced edge set. For the anigrafs with vertices {A, B ... F}, none of the four vertices {C, D, E, F} had the same edges. Thus, only A and B are identical to A′, B′ , and the similarity $S(2)$ = 1/3. For the {P, Q ... U} pairs, vertices P, P′ and Q, Q′ have the same edges, and so does vertices T,T′. Hence $S(2)$ = 1/2. Structural similarity for anigrafs is thus dictated by vertex similarities and only indirectly by the edge sets.

8.2 Similarity versus Outcomes

As the similarity between two anigrafs decreases, so will the percent of outcomes that are the same. To explore this relationship in detail, generate a set of random graphs {G_i} with n vertices and with edges chosen with probability 1/2. For each of these graphs we now revise a fraction, f, of the edges to create a new set of graphs {H_i}. For large n, we find that the chance of G_i and H_i having the same Condorcet winner is inversely proportional to the structural dissimilarity between H_i and G_i. This relationship is shown by the diagonal line in figure 8.4.

One might ask whether all tally procedures will generate a result similar to figure 8.4. The answer is no. If the vertex with maximum weight is always chosen as the winner, then

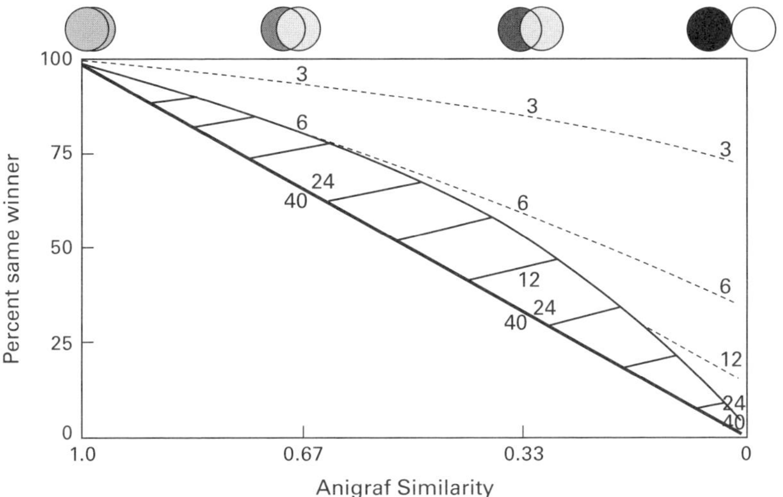

Figure 8.4

Similarity of anigrafs, Kd=2; number of vertices as labeled.

all curves will be flat at 100 percent regardless of their similarity. The Condorcet tally is special in that it incorporates information about second or higher order preferences, which are also used as the basis for defining similarity. Knowledge depth (Kd) now becomes an important parameter in relating similarities and outcomes. If the knowledge depth is increased, the results of figure 8.4 will change, especially for small graphs. The reason is because deeper preference rankings will include relations for vertices previously chosen to have fixed edges to their neighbors. Consider the right pair of anigrafs in figure 8.3. If the knowledge depth is increased from two to full, then the preference order for P changes from [P > R > Q, S, T, U] to [P > R> Q~S > T~U]. Thus, the Condorcet (and Borda*) tallies will change, subject to the particular way the second anigraf is augmented. The exception will be for graphs with diameter two.

8.3   Coordination and Diversity

Our objective is to create a diverse group of anigrafs that still have some unity through shared goals and preferences. But figure 8.4 presents a dilemma: as graphical forms become less similar and more diverse, the odds for a pair of random anigrafs having the same winner decrease proportionately. How can we optimize diversity and similarity at the same time?

Although the linear relation between similarity and common outcomes implies only one underlying variable, this is not the case. There is another variable obscured by the deceivingly simple figure 8.4. Absent is just how tightly the percent of common winners are distributed about the mean for any given similarity. For example, a set of anigraf quintuples will always agree on winners; thus, the variance in common winners is zero. Likewise, for a collection of (large) random anigrafs, the probability of any pair having the same winner will be zero, and, again, the variance about the (zero) mean will be zero. Between these two extremes, there will be a distribution of the percentage of common winners for any similarity index. This relationship is shown by the crosshatched envelope in figure 8.4 for anigrafs of size 24. We now use this variation to advantage.

Let us repeat the initial experiment where G_i and H_i have different graphical forms, but now also compare winners with a third or fourth variation, H_i or J_i, of G_i, where all variations have the same similarity measure with respect to G_i. Using figure 8.4, we can now estimate the probability that at least one of the pairings with G_i, namely G_i, H_i or G_i, J_i, etc., will have the same winner. This pairing probability gives an indication of the strength of group unity. (Pairs H_i, J_i, etc., are correlated, and therefore excluded from the estimate.) Multiplying this

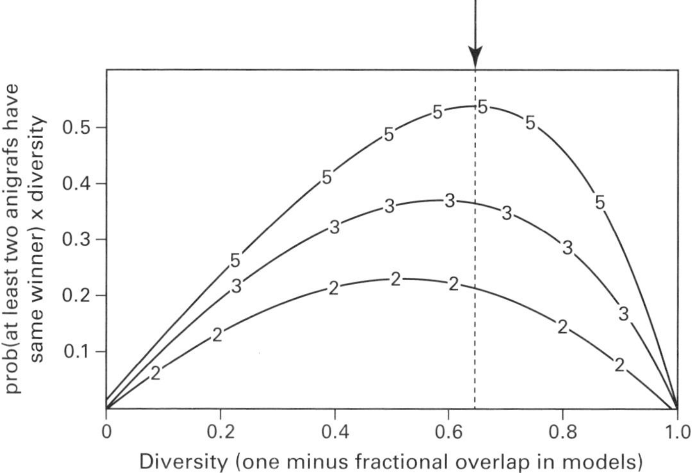

Figure 8.5

Relation between diversity and winner probabilities.

probability-of-unity times the dissimilarity of form, namely $(1 - S(2))$, gives a measure of "coordinated diversity." Maximizing this product yields the optimal balance between coordination, or group unity, and diversity.

Figure 8.5 shows this coordinated-diversity product for small groups of size 2, 3, and 5. The maximum moves toward a 2/3 dissimilarity of form, based on simulations using random graphs with edge probability of 1/2. Thus an estimate for the optimal diversity among members of an anigraf population is when the similarity between pairs is about 1/3.

8.4   Clique Formation

Within any population of anigrafs, benefits for some can be gained by forming subgroups where members agree to vote as a block. Obviously, members of such groups will resemble one another and will have strong similarities in preferences and choice of winners. What role will the anigraf form play in the likelihood that a set of anigrafs will be able to share common preferences?

Definition: A clique of size m is a set of anigrafs of size n where each of the m members shares at least one preference relation with at least one other member of the clique.

In other words, if any two members of a clique are chosen, there will be at least one goal for each anigraf that is the same for both, and, for knowledge depth = 2, there will be the same alternate goals or choices (i.e., the neighbors to that vertex in the anigraf are the same).

Now consider random anigrafs of size 12 with edge probability of 1/2. From figure 8.4, we know that there is roughly a 10 percent chance that any two anigrafs will have on average one agent or mental organism with the same preference orders. In such a population, with a small likelihood of even one shared vertex between members, the clique size will be small (relative to the population size), and conversely, the number of cliques will be large. However, if we were to increase the edge probability toward one, then the anigrafs will become increasingly more similar, eventually all being identical, forming one large clique. This effect of edge probability on expected clique size is shown in figure 8.6 for a population of twenty different anigrafs.

In the right panel, for edge probability 1/2, almost all individual anigrafs are isolated, and the maximum clique size is very small. As edge probability goes to one, clique size increases to the size of the population of twenty. Along with this trend, as shown in the left panel, the average number of cliques with at least two members will increase, except at the limiting value when the network is completely self-connected.

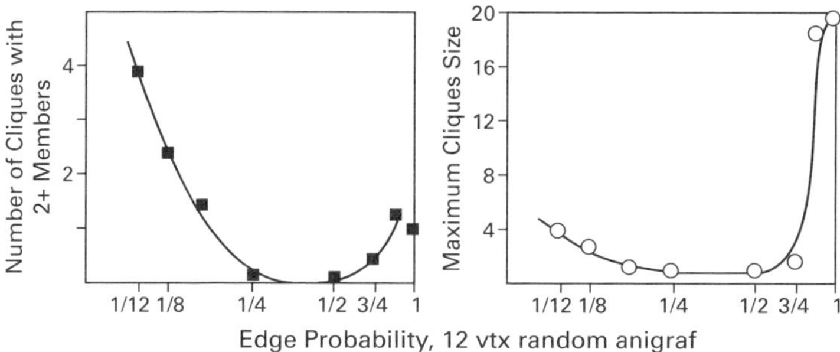

Figure 8.6

Relation between edge probability and clique size (right) and average number of cliques with at least two members (left).

The surprise is when anigraf edges become sparse, with edge probability going toward zero. Now maximum clique size increases again, and more and more cliques appear with at least two members. This multi-modal distribution is a key to the formation of social networks: If a unified population is to consist of dissimilar "random" anigrafs, then the division of this population into cliques of any significant size will require sparse anigraf designs. But sparse anigrafs begin to resemble trees.

8.5   Social Networks

Roughly speaking, a social network is a network of anigrafs, with different species compatible through some similarities in structure. The members of the network are typically very small, generally having only a dozen nodes or fewer. (Larger anigrafs—having more than twelve alternatives—present difficult choice sets.) We assume that links between the anigraf members of the network are sparse, with these links favoring vertices that have similar choices. In particular, the first link

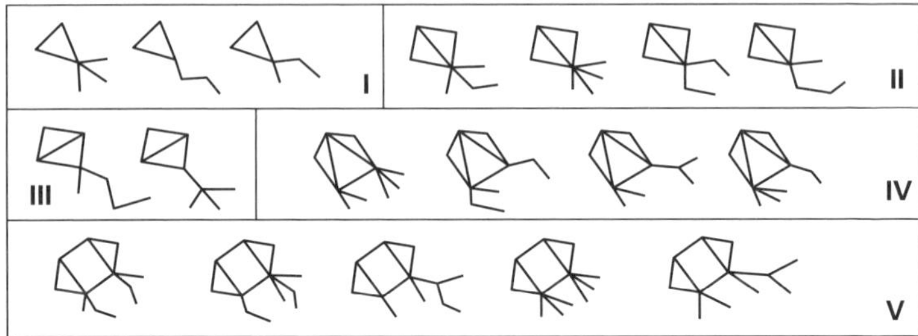

Figure 8.7

Some of the more common clique members.

is given preference. If subsequent neighbors to these primary links are chosen among neighbors of the first choices for a link, then that network will be scale-free (Richards and Wormald 2015; Barabasi and Albert 1999). Furthermore, the network will be loosely connected, with the density of vertices between cliques being much less than the density of the similar vertices between constituent anigraf members.

To give an example, first fix the knowledge depth of the anigrafs at two, considering only neighbors to vertices when evaluating similarity. Now generate a population of fifty anigrafs having 6, 8, 10, and 12 nodes. Each of these has one fixed subgraph, such as a triangle, a square with one diagonal, a pentagon with one vertex of degree four, etc. (See figure 8.7.) Attached to this fixed subgraph, or "body," are limbs and appendages. These attachments are generated as random trees, always rooted at the same vertex of the body. Both the fixed body subgraphs and their attached random trees have the same number of vertices—namely 3, 4, 5, and 6, respectively, for the different size anigrafs. The design of the body subgraphs ensures that cliques will be formed among the different species of anigrafs; the random tree attachments provide the basis for

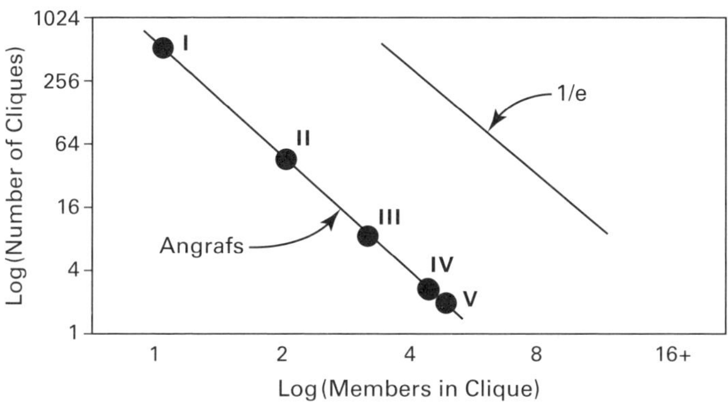

Figure 8.8

Number of cliques for anigrafs illustrated in figure 8.7.

similarities between different species, thus forming the basis for a social network.

Figure 8.8 shows the number of members (cliques) of any anigraf size that are formed when several populations are generated, and when we require four similar vertices between two or more members of a clique. There are a large number of small cliques, and few very large cliques. It is the treelike form of the limbs of the anigrafs that leads to this relation. Vertices of degree one are easiest to link, then those of degree two, etc. Therefore, because their tree is larger, larger anigrafs will be more likely to have vertices of degree one, and, consequently, will find more clique members. This leads to a slope of the relation between clique size and number of roughly 1/e, where e = 2.718.

Although each clique by itself has a social structure reflected in its clustering properties (on average 80 percent of that of the comparable complete graph), this is not the social network of interest here. Different species cannot exist independently of others. How will a society of anigrafs of different types and sizes relate to one another? These relations specify a much larger, complex anigraf that characterizes the nature of a higher-order social structure. For our primitive anigraf collection, the answer lies in how the various cliques are linked and interconnected. As previously explained, the most common affinities are among cliques with the smallest anigrafs. This is true in our example because there are very few different tree designs for their limbs. In contrast, larger anigrafs, with six or more vertex trees, will have many forms for their limbs. Most of these forms will be quite different from those of the very small anigrafs with only three-vertex trees. Links in the social network will thus tend to be between anigrafs of the same size or between anigrafs of slightly different sizes.

An intriguing problem is as anigraf size grows how will an increased density of similarity links within a clique narrow the ability to relate to anigrafs of either larger or smaller size—or to anigrafs with different basic designs. Is there one anigraf form that can optimize relationships both within and between cliques? Anigraf architectures have a hierarchical aggregation of information leading to decision-making. At the lowest levels, we have entities that select among possible limb movements; at high levels there are planners and brokers. At each of these levels, the architecture is similar, relying on graphical forms that capture similarity relations among alternatives. These relational graphs are small, generally having only ten nodes or fewer. The possible variations are of order 10^7. However, constraints on anigraf designs will reduce the options by several orders of magnitude. How might these small graphs be aggregated into a larger whole?

One possibility is if some alternatives appear as participants in several different relational graphs. For example, if an alternative is a cover in one of ten small graphs, each with ten

Box 8.1
A Daemon's Language

Communication between daemons implies communications among neural modules. Because each module needs to understand another's perspective, the implication is that the modules have a language in common. This means not only a shared grammar, but also a shared vocabulary. Simple firing activity of neurons seems an unlikely method to fulfill these conditions. A solution to this problem would be for each module to have a portion of its activity devoted to the same task as executed by adjacent modules. Fortunately, this possibility receives support from single unit studies. In perception, for example, we see that the area MT responds to more than motion, and V4 neurons respond to color as well as other sensory attributes (wikipedia.org/visual_cortex). Presumably, the same kind of diversity is also seen in the motor areas. Note that this attribute is exactly what appears in the tally machine. Hence the tally machine "solves" the problem of a common language between similar actions (or percepts). To implement this option, we would need only a set of overlapping discs, each with their own attributes, with the overlap being the shared part of the tally machine. In this manner, each neural module would have cells responding to a modality other than its primary one. Note also that this model suggests that the weights shared among alternatives are not reciprocal. Indeed, one module might contribute 30 percent of its total weight, whereas a bigger module might be contributing only 10 percent. Simulations are needed to explore this option in the tally process.

nodes, then that alternative will have a vertex degree of 100. This possibility would provide support for a multiscale social network (Kasturirangan 2010). To elaborate, note that the set of action verbs constituting one important class of alternatives will have a power law distribution, as proposed by Zipf (1932). The anigraf architectures will then have a vertex degree distribution typical of social networks (Barabasi-Albert 2010).

Part IV

Metagrafs

Metagrafs are relationships between anigraf models. Examples would include how one model may be transformed or decomposed into another, or similarity relations between graphs. The simplest such relationship is the identity, when one graph is the same as another, but the nodes have different meanings in different contexts. Metagraphs have the potential to point out new anigraf designs by revealing different classes of graphical forms (e.g., see Gunkel's networks). For advanced creatures, a metagraf can also provide an anigraf with self-induced insights. When one graph has the same form as another, but each are in different contexts, with different labels on nodes, then an analogy is created that offers insight into the nature of the new context. Metagrafs are thus operations on graphs. However, unlike anigrafs, metagrafs are not separate entities.

9.1   Representational Forms

An obvious metagraf transformation is when one rule is used to construct a graph, and then the graph is grown by adding more copies of the precursor graph (Wolfram 2002). Simple examples of metagraf transformation, where the class of the graph remains the same, include: extending a chain graph, enlarging a ring, and enlarging a series of regular graphs.

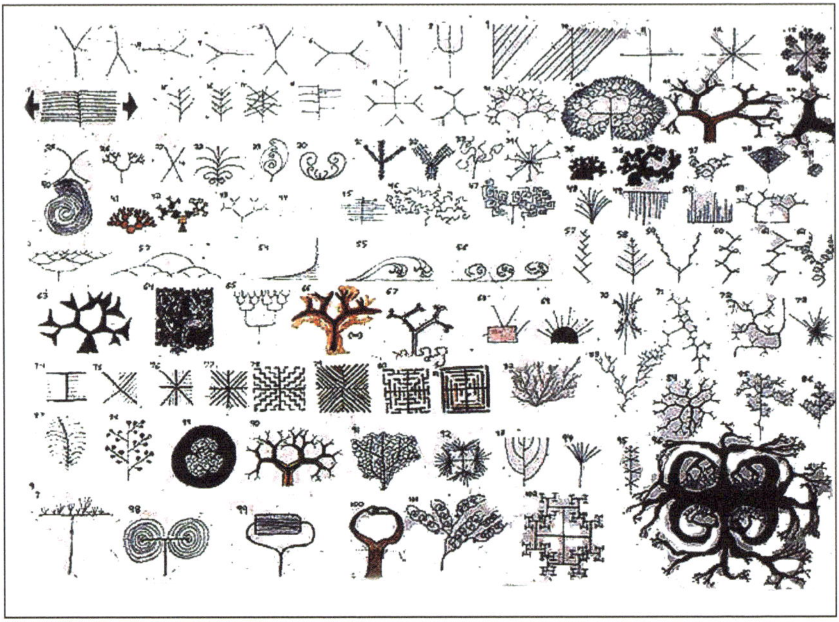

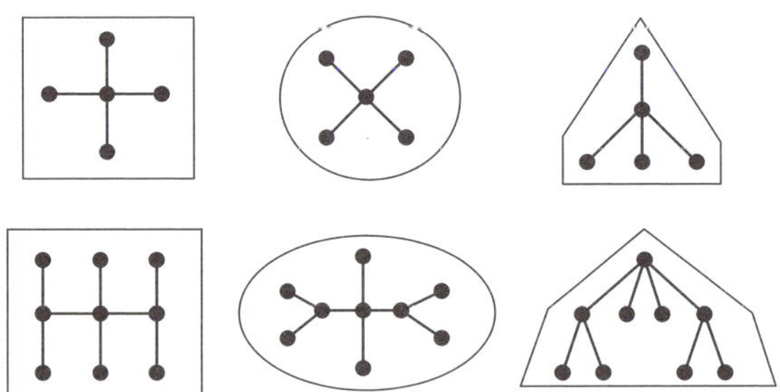

Figure 9.1

Various network types conceived by P. Gunkel in *Ideonomy*.

Figure 9.2

Picking a different root or frame for a graph can change its interpretation. (Note that there are only two graphical forms—one for each row.)

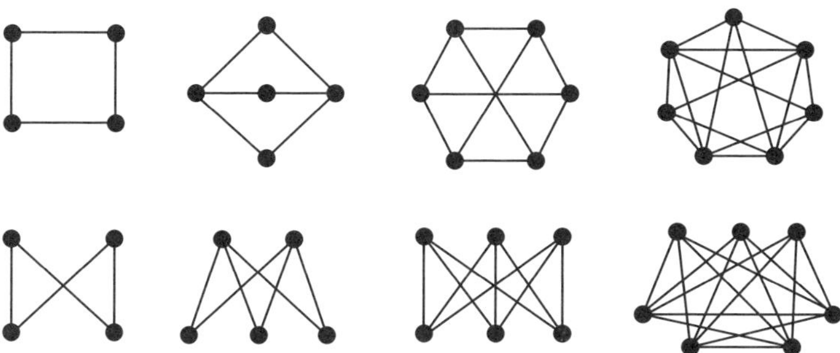

Figure 9.3

Each column is the same graph, but reconfigured to emphasize different properties.

A more interesting transformation, however, is a change in context or in the "root"—i.e., main node—of a graph. Figure 9.2 illustrates such a transformation, by changing both the root and the frame for a graph. Most people do not immediately see that there are only two different graphs in this figure. Figure 9.3 illustrates another example. Here the lower row duplicates the graphs in the upper row. The transformation is simply a reconfiguration of how the graph is displayed. Although both examples are rather trivial, they emphasize what is called "framing" in persuasive arguments. Franklin Roosevelt used this technique frequently: by changing the context, a message otherwise difficult to grasp can be conveyed.

Such reconfigurations can be fundamental to how anigrafs evolve. Although the underlying structure of the anigraf remains the same, as more nodes and precursor graphs are added in different contexts, the anigraf form will likely change. However, there will remain an underlying species similarity, with the more mature forms exhibiting different structures.

Metagraf operations on graphs may also have a dynamic that anigraf models do not. Transformations along paths imply movement; similarities between pairs are stationary relationships. If there is a dynamic to a relationship between anigraf models, then this dynamic takes on a metagrafical form in its own right, with model transformations that continue except at a terminus. Changing parameterizations, root nodes, or foci can all cause a restructuring of dynamic form. Elements of events seen before as positives can become negatives. Recasting forms in a fixed context, as in the complements of figure 9.5, or altering the context to change perspective on a form, as in figure 9.1 are cognitive operations that reshape thought. We see the prevalent use of analogs, or of changes in depictions, or new perspectives gained through a change in context. What laws govern this space of cognitive inquiries? More importantly, by understanding the constraint on cognitive operations, can shackles be broken to gain freedom for greater creativity?

9.2   Complementarity

Models of models are highly cognitive artifacts. Like trees, certain parameters and transformations tend to reoccur, thereby specifying identical or similar metagrafs. The spectrum of love to hate, friend to foe, good to evil are examples. Each element lies at the opposite extreme of a signed parameterization. With most terminal paths through relational spaces comes the notion of opposites, or, more generally, complementarity. These are pairs about an origin or axis that are not necessarily at the extremes, but are in some sense objects located at symmetric positions in the space. The mapping of complements is thus a metagraf.

For connected graphs, complements can be specified precisely, using a slight revision of the graph-theoretic version:

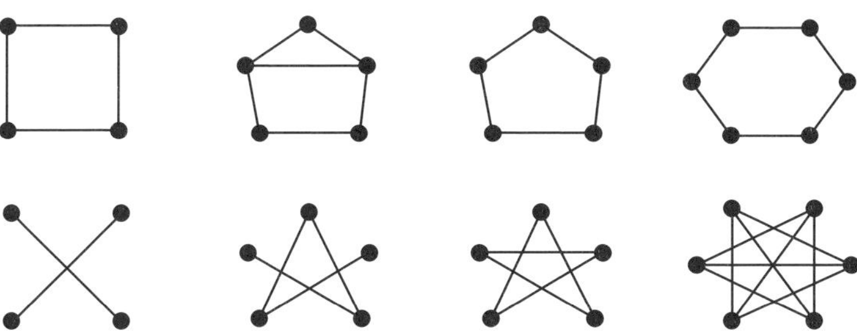

Figure 9.4

Each graph in the lower row is the complement of one in the upper row.

Definition: The (connected) complement $\underline{G}_n$ of the graph G_n is the connected edge set of K_n obtained by removing the edges of G_n from K_n, subject to each vertex in G_n or $\underline{G}_n$ being at least degree one.

Figure 9.4 illustrates. The upper and lower rows are complements of each other. In the third case, $\underline{G}_n$ and G_n' are self-complements. For these ring graphs, note that their complements (without unfolding) appear to belong to a class of "starlike" graphs. The complementing operation on rings in this context has produced a new set of similar graphs that belong to the metagraf complement of the ring metagraf. It is very unlikely that anigrafs in complementary sets can "understand" or relate to one another. However, would such anigrafs still create realistic models of their worlds? Can we have models based on dissimilarities, rather than similarities (Hawking and Mlodinow 2010)?

9.3   Decompositions

Anigafs are focused on relationships, not the form of the objects themselves. Rather, the nature of the objects is assumed

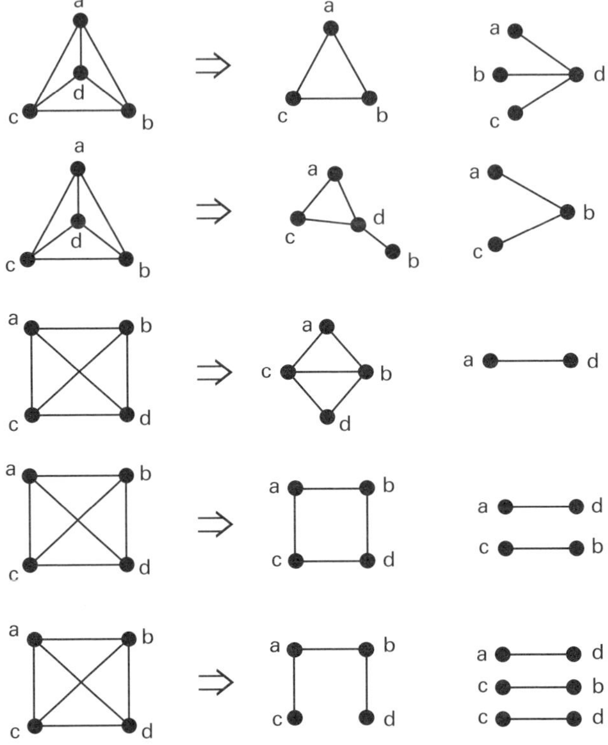

Figure 9.5

Pairs to the right of the arrow show interpretable decompositions of the graph in the left column.

to set the context, thereby influencing the observed relationships. The basic idea is best illustrated with K_4. There are five different kinds of subgraphs that result when an edge is removed from K_4. These, and their residuals, are shown at the right in figure 9.5. For clarity, two depictions of K_4 are used, one having a triangular outline, the other a square. At the top, we have $K_4 = K_3 + t_3$. Indeed, for any K_n, it is obvious that $K_n = K_{n-1} + t_{n-1}$. Hence any K_n can be decomposed into a (non-unique) set of trees. In the third row of the figure, $K_4 = R_4$ + $\{g_i\}$, where g_i in this case are two versions of K_2. Still more possibilities are given in the remaining rows.

If graphs are viewed as patterns, there is nothing especially striking about these decompositions. For example, a ring and two edge segments with nodes seem quite distinctively different and hard to "relate" to one another. Perhaps the strongest relationships are the top two triangular forms in the second column, and the two square rings below. Ideally, we would like the subgraph decompositions of K_n to have compelling relationships.

9.3.1 Hybrid Decompositions

Consider a novel decomposition of the complete graph K_5. Let K_n be divided initially into complementary pairs having as nearly as possible the same number of edges, and thus creating an "equal" partitioning. Note that in this case the first decomposition yields two identical graphs (a pentagon) set in different frames. Now proceed to manipulate these pairs by adding an edge to one, and removing an edge from the other. Figure 9.6 shows this result for K_5. Note that for the anigraf in this depiction, the initial split is very compelling. The transitions seem to follow rather well. The entire structure is a candidate for the metagrafical organization of K_5, seen as a decomposition of connected subgraphs of order 5. (The lower orders follow trivially.) Perhaps one kind of "beauty" associated with K_5 is revealed in these subgraph relationships. Of interest is that all of these decompositions contain the chain C_4 as a subgraph.

9.3.2 Ramsey-like Decompositions

Given n objects under consideration, let the edges of K_n represent all possible bidirectional (nonmetric) relationships. We

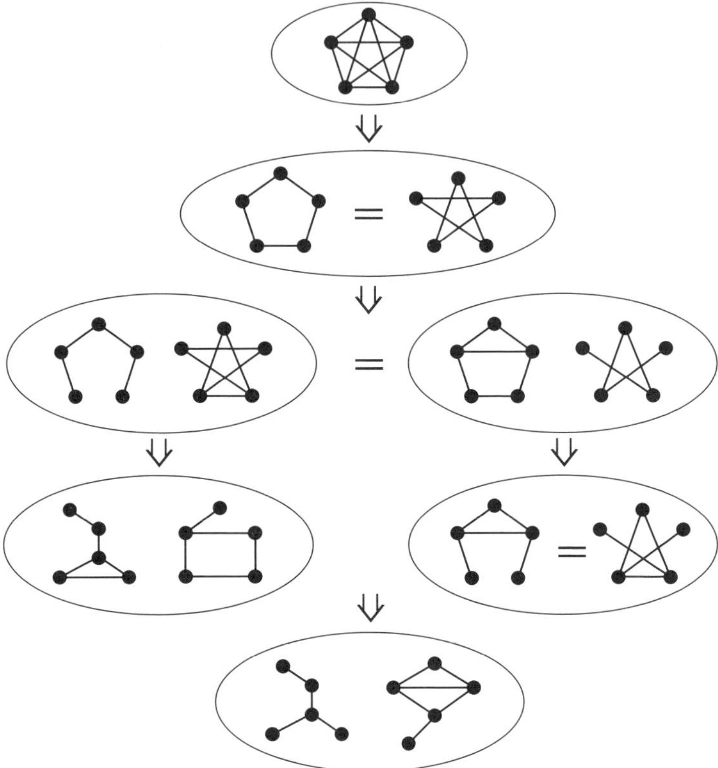

Figure 9.6

Decompositions of the subgraphs of K_5.

now ask, given any decomposition of these edges into two sets, is there a connected subgraph always guaranteed to be present in at least one of the decompositions? If so, then those subgraphs present for K_n but not appearing in decompositions of K_{n-1} will be designated the "atoms" for K_n. Clearly one element of a chain, namely C_2, is the simplest atom, but does not qualify as an atom for K_n, $n > 2$. Therefore, the atoms of special interest are the "largest" for any K_n—those that appear first for K_n, but not for all the decompositions of K_{n-1}.

Definition: The "atoms" $å_n$ for K_n are the connected subgraphs that are present in at least one member of all binary decompositions of the edge sets of K_n, but are not present in all such decompositions of K_{n-1}.

Not surprisingly, most of these atoms have appeared throughout the text.

9.4 Evolutions

When a graph evolves or is altered, often the old frame seems inappropriate. In figure 9.7 for example, the "+" graph can be augmented by adding a diagonal edge "xy." But when this happens, the original parameterizations within the frame are broken. Similarly, in the "T" graph, edge "ad" appears an unreasonable leap in the context, obviating the linear a, b, c, d

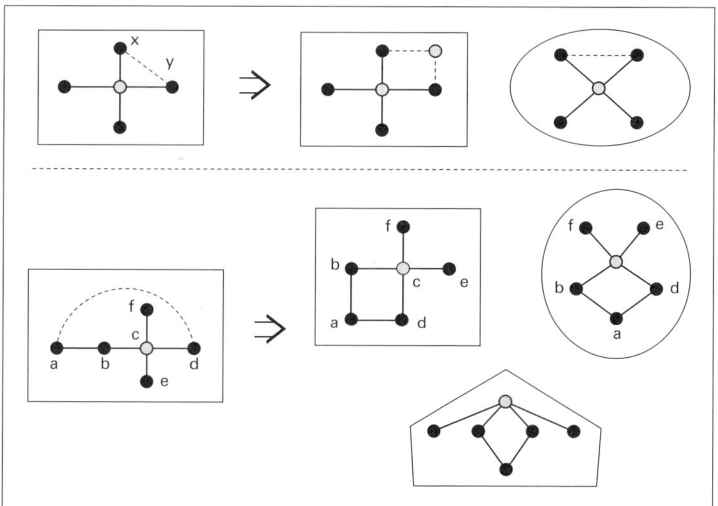

Figure 9.7

Two graphs are altered by adding an edge, which leads to a revision of their frames.

parameterization. Two "solutions" are considered: the first strives to retain the original frame; the second keeps the graphical form, but changes the frame, leading to a revision in context. In the case of a "+" with an added diagonal edge, the graph cannot be fully aligned with a rectangular frame. If the frame is to be preserved, a node with two edges properly aligned is "forced" to be added. However, if the rectangular frame is changed to a circle, then we have a satisfactory coordinate frame embedding of the graph, as indicated at the far right. Alternatively, we could consider a rooted dendrogram; but then the added edge (dashed) has no relevance in this particular tree.

In a similar manner, when an extreme edge "ad" is added to a "T" graph, the most compelling "solution" is to realign the edges to create a square with a forked tail. This can be embedded nicely in a new ellipse or rectangular frame, depending upon the meaning of the tail. Again, one could also grasp either one of these new configurations by the root to accommodate the dendrogram frame. These operations support creative discovery because a new metagraf structure is needed to specify the relationships between models.

9.5   From Decomposition to Composition

As anigraf systems evolve, a variety of forms will result, depending upon internal and external constraints. This evolution will likely use a combination of merging, splitting, and framing operations. Wolfram (2002) provides several examples using a small basic graph with rules for combining and augmenting the new, larger forms. Simulations will probably be the best tools for understanding network evolution (Epstein 2007). However, to date, work on social networks has failed to reveal any underlying basic component of the network (Macindoe 2010; Richards and Wormald 2015). The implication is thus that composition, unlike decomposition, will be largely intractable. This conclusion is reinforced when one realizes how networks evolve. If they evolve by preferential attachments, say to similar or related nodes, then one first needs to understand the domain of related alternatives, not the actual graphical form. Predicting similarities among alternatives is not a formal endeavor; rather one might have to use the empirical data such as that provided by WordNet.

Decompositions, on the other hand, are tractable, as shown in the previous sections. Rules for breaking patterns down will typically lead to more basic graphs, which can appear as components of quite different larger graphs. With techniques such as nonmetric multidimensional scaling (Shepard 1962) or trajectory mapping (Richards and Koenderink 1995), which enable the experimenter to recover similarity relations among alternatives, the substructure of anigraf designs may thus be accessible.

Epilogue

Superficially, anigrafs may appear to be simple enhancements of Braitenberg's vehicles, expanded to include aspects of cognition. This similarity however should not obscure the fact that anigraf mechanisms are quite different from vehicle servo-feedback systems. These differences in design favor mimicking behaviors set in quite different contexts. Anigrafs explore the consequences of viewing living creatures as a society of agents, with beliefs and preferences captured by relationships of a graph. These differences open new windows to understanding mind, revealing computational complexities, relationships, and structures not previously considered.

The reality of "mind" nevertheless remains elusive. Considering mental events as part of a society of mental organisms can sharpen our view. Within a society, useful communications and cooperativities result in behaviors related by models and customs that are shared by a community. There is a sense that one's actions are not simply a hodgepodge of reflexive activities. This implies an underlying set of constraints that enforce a nonrandom structure for mental activities, with some underlying neural machinery. Anigrafs address this problem. To sharpen the distinction between mind and mechanism, consider that: "mind is to a brain as time is to a clock" (Cowan 1998). Clocks have many different forms such as atomic, spring wound, and pendular. Yet all serve the same function—namely, to keep time. Similarly, just as we have many forms for clocks, so brains and their components may differ widely—especially as silicon and carbon-based machines are further developed. Despite these differences, all are seen as "controlling" behaviors of encapsulated entities. It is the observation of this behavior that gives brains meaning. Just as clocks are meaningless without a notion of time, so brains need a notion of mental activities that observers (including the creature itself) can recognize and typically utilize.

Although anigrafs give only a glimpse of the very lowest levels of cognition, the ideas may provide a platform for models of higher cognitive capabilities. For example, the understanding of ethics and morality seems within reach because these are social constraints likely to be shared by an anigraf society. Much more difficult would be to model aspects of beauty, such as a painting by Rembrandt, a musical composition by Beethoven, a poem by Frost. Is it possible to model why a sunset is so special, or a butterfly, or a flower? Is it the relations among what we consider the "parts" of the object or events we observe? Or perhaps the transformations or group-theoretic structures are the key. If so, then we will need to allow the more advanced anigrafs to recognize and manipulate their own transformations and relational forms. Metagrafs then becomes a new, viable route for study.

Box 10.1
On Consciousness

Recently, G. Tononi (2008) outlined a theory of consciousness as integrated information. His novel proposal has received some unexpected support (Koch, 2012). Surprisingly, the three important aspects of Tononi's theory have counterparts that

Box 10.1
(continued)

appear in the anigraf architecture. The three components are: (i) there is an uncertainty measure of the quantity of consciousness (considered here as "the strength" or more appropriately "vividness"); (ii) the informational relationships among the constituents (agents) are seen as relations between probability distributions; and (iii) the conscious experience is captured geometrically and is more than the sum of the states of the components.

For anigrafs, integrated information is an outcome aggregated by a tally machine. Although there is no specific information measure, there is an obvious link between the uncertainty reduction of the varied desires of the daemons. If all agree on an outcome (as indicated by the lack of local- or top-cycles), then uncertainty is nil. Such a result would be seen pictorially in the phase plots as homogeneous regions without texture.

In the second part of Tononi's proposal there are informational relationships among the agents/daemon's desires that are cast as probability distributions. In anigrafs, these are captured by the similarity graph of alternatives. This immediately creates a geometric space among constituents analogous to Tononi's figure 3. In anigrafs, these overlapping preference relationships also serve the important function of a "language" shared by daemons. The degree of overlap could also provide a basis for joint probability distributions, as required by Tononi. This information cannot be recovered from the tally operation, which is Tononi's third condition.

Bibliography

Amaari, S., and M. Arbib. 1977. "Competition and cooperation in neural nets." In *Systems Neuroscience*, edited by J. Metzler, 119–165. New York: Academic Press.

Arbib, M. A. 1964. "Cybernetics." In *Brains, Machines, and Mathematics*, 93–105. New York: McGraw-Hill.

Arrow, K. J. 1951, 1963. *Social Choice and Individual Values*. 2nd ed. New Haven: Yale University Press.

Axelrod, R., and D. Dean. 1988. "Further evolution of cooperation." *Science* 242:1385.

Barabasi, A.-L. 2002. *Linked: The New Science of Networks*. Cambridge, MA: Perseus.

Barabasi, A.-L., and R. Albert. 1999. "Emergence of scaling in random networks." *Science* 286:509–511.

Beinecke, L. W., and R. J. Wilson. 1997. *Graphical Connections*. Oxford: Clarendon Press.

Berg, H. C. 1996. "Touring Machines." *Current Biology* 6:624.

Black, D. 1958. *The Theory of Committees and Elections*. Cambridge: Cambridge University Press.

Bond, A. 1999. "Describing behavioral states using a system model of the primate brain." *American Journal of Primatology* 49:315–338.

Borda, J–C. 1781. "Memoire sur les elections au Scrutin." *Histoire de l'Academie Royale des Sciences*.

Braitenberg, V. 1984. *Vehicles: Experiments in Synthetic Psychology*. Cambridge, MA: MIT Press.

Brinkmann, G., and B. D. McKay. 2002. "Posets on up to 16 points." *Order* 19:147–178.

Conant, R., and R. Ashby. 1970. "Every good regulator of a system must be a model of that system." *International Journal of Systems Science* 1, no. 2:89–97.

Condorcet, Marquis de. 1785. *Essai sur l'application de l'analyse a la probabilite des decisions rendues a la probabilitie des voix*. Paris: De l'imprimerie royale.

Couzin, I., J. Krause, N. Franks, and S. Levin. 2005. "Effective leadership and decision-making in animal groups on the move." *Nature* 433:513–516.

Cowan, H. J. 1958. *Time and Its Measurement*. Cleveland: World.

Crick, F., and C. Koch. 2003. "A framework for consciousness." *Nature Neuroscience* 6:119–126.

Darwin, C. R. 1859. *On the Origin of Species by Means of Natural Selection*. London: John Murray.

Davis, J., A. Bobick, and W. Richards. 2000. "Categorical representation and recognition of oscillatory motion patterns." IEEE Conference on Computer Vision and Pattern Recognition, 628–635.

Dorogovtsev, S. N., and J. F. Mendes. 2003. *Evolution of Networks*. Oxford: Oxford University Press.

Epstein, J. M. 2006. *Generative Social Science: Studies in Agent-Based Computational Modeling*. Princeton: Princeton University Press.

Feldman, J. 2014. "Bayesian models of perceptual organization." In *Oxford Handbook of Perceptual Organization*. Oxford: Oxford University Press.

Glimcher, P. 2010. *Foundations of Neuroeconomic Analysis*. Oxford: Oxford University Press.

Gould, S. J., and N. Eldredge. 1977. "Punctuated equilibria: the tempo and mode of evolution reconsidered." *Paleobiology* 3:115–151.

Greene, P. 1962. "On looking for neural networks and 'cell assemblies' that underlie behavior: a mathematical model." *Bulletin of Mathematical Biophysics*.

Gunkel, P. 2000. *Ideonomy*. Available at http://ideonomy.mit.edu.

Harary, F. 1969. *Graph Theory*. Reading, MA: Addison-Wesley.

Hawking, S., and L. Mlodinow. 2010. *The Grand Design*. New York: Bantam Books.

Hayek, F.A. 1952. *The Sensory Order*. Chicago: University of Chicago Press.

Heider, F., and M. Simmel. 1944. "An experimental study of apparent behavior." *American Journal of Psychology* 57:243–259.

Hildebrand, M. 1967. "Analysis of symmetrical gaits in tetrapods. *American Journal of Physical Anthropology* 26:119–130.

Hoffman, D. D. 2008. "Conscious realism and the mind-body problem." *Mind & Matter* 6:87–121.

Hopfield, J. J., and D. Tank. 1986. "Computing with neural circuits: a model." *Science* 253:625–633.

Huntington, S. P. 1996. *The Clash of Civilizations and the Remaking of World Order*. New York: Simon & Schuster.

Kasturirangan, R. 1999. "Multiple scales in small world networks." MIT Artifical Intelligence Memo 1663.

Kamler, C. 2008. "The Dodgson ranking and the Borda Count: A binary comparison." *Mathematical Social Sciences* 48: 103–108.

Koch, C. 2012 *Consciousness: Confessions of a Romantic Reductionist*, The MIT Press

Li, T-Y, and J. A. Yorke. 1975. "Period three implies chaos." *American Mathematical Monthly* 82:985–992.

Lieberman, E., C. Heuert, and M. A. Nowak. 2005. "Evolutionary dynamics on graphs." *Nature* 433:312–316.

Lorentz, E. N. 1963. "Deterministic non-periodic flow." *Journal of Atmosheric Sciences* 20:130–141.

Lorenz, K. 1982. *Foundations of Ethology*. New York: Simon & Schuster.

Maas, W. 2000. "On the computational power of winner-take-all." *Neural Computing* 12:2579–2635.

MacKay, D. M. 1969. *Information, Mechanism, and Meaning*. Cambridge, MA: MIT Press.

MacMahon, T. A. 1984. *Muscles, Reflexes, and Locomotion*. Princeton: Princeton University Press.

Macindoe, O., and W. Richards. 2010. Graph comparison using fine structure analysis. Social Computing. IEEE Second International Conference, 193–200.

Mandelbrot, B. B. 1982. *The Fractal Geometry of Nature*. New York: Freeman & Co.

Marler, P. 1987. "Learning by instinct." *Scientific American* 256, no. 1.

Marr, D. (1970) A theory for cerebral neocortex. *Proceedings of the Royal Society, London B* 176:161–234.

Minsky, M. 1986. *Society of Mind*. New York: Simon & Schuster.

Moon, J. W. 1968. *Topics on Tournaments*. New York: Holt, Rinehart & Winston.

Mowshowitz, A., and M. Dehmer. 2012. "Entropy and the complexity of graphs revisited." *Entropy* 14:559–570.

Nauta, W., and M. Feirtag. 1986. *Fundamental Neuroanatomy*. New York: Freeman.

Newman, M. E. J., A.-L. Barabasi, and D. Watts, eds. 2005. *Structure and Dynamics of Complex Networks*. Princeton: Princeton University Press.

O'Regan, J. K. 1992. "Solving the 'real' mysteries of visual perception: The world as an outside memory." *Canadian Journal of Psychology* 46:461–488.

Page, S. E. 2007. *How the Power of Diversity Creates Better Groups, Firms, Schools and Societies*. Princeton: Princeton University Press.

Palmer, E. M. 1985. *Graphical Evolution*. Malden, MA: Wiley & Sons.Paynter, H. M. 1961. Analysis and Design of Engineering Systems (class notes for subject 2.751). Cambridge, MA: MIT Press.

Pentland, A. 2008. *Honest Signals: How They Shape Our World*. Cambridge, MA: MIT Press.

Poston, T., and I. Stewart. 1978. *Catastrophe Theory and Its Applications*. London: Pitman Publishing.

Prelec, D., and H. S. Seung. 2012. An Algorithm That Finds Truth Even If Most People Are Wrong. In press.

Purtell, T. J. 2004. Graph Model for Artificial Intelligence. Advanced Undergraduate Project; Department of Electrical Engineering and Computer Science; MIT.

Raibert, M. 1988. "Balance and symmetry in running." In *Natural Computation*, ed. W. Richards. Cambridge, MA: MIT Press.

Ramachandran, V. S. 2000. Mirror neurons and imitation learning as the driving force behind 'the great leap forward. in human evolution. *Edge*. http://edge.org/3rd_culture/ramachandran/ramachandran_p1 .html.

Rashevsky, N. 1955. "Life, information theory, and topology." *Bulletin of Mathematical Biophysics* 17:229–235.

Read, R. C., and R. J. Wilson. 1998. *An Atlas of Graphs*. Oxford: Oxford University Press.

Resnick, M. 1994. *Turtles, Termites, and Traffic Jams: Explorations in Massively Parallel Microworlds*. Cambridge, MA: MIT Press.

Richards, W., and J. J. Koenderink. 1995. "Trajectory Mapping: A new nonmetric scaling technique." *Perception* 24, no. 11:1315–1331.

Richards, D., B. D. McKay, and W. Richards. 1998. "Collective Choice and mutual knowledge structures." *Advances in Complex Systems* 1:221–236.

Richards, W., B. D. McKay, and D. Richards. 2002. "The Probability of Collective Choice with Shared Knowledge Structures." *Journal of Mathematical Psychology* 46:338–351.

Richards, W., H. R. Wilson, and M. A. Sommers. 1994. "Chaos in Percepts?" *Biological Cybernetics* 70:345–349.

Richards, W., and N. Wormald. 2015. "The evolution and structure of social networks." *Network Science*. (in press)

Rizzolatti, G., and L. Craighero. 2004. "The mirror-neuron system." *Annual Review of Neuroscience* 27:169–192.

Rossi, R.A., J. Neville, B. Gallagher, and K. Henderson. 2011. "Modeling dynamics behavior in large evolving graphs." LLNL-TR-514271. (In *Association of Computing Machinery*, 2013.)

Runkel, P. J. 1956. "Cognitive similarity in facilitating communication." *Sociometry* 19:178–191.

Saari, D. 1994. *Basic Geometry of Voting*. Berlin: Springer-Verlag.

Saaty, T. L., and L. G. Vargas. 1984. "Inconsistency and rank preservation." *Journal of Mathematical Psychology* 28:205–214.

Searle, J. R. 1983. *Intentionality*. Cambridge: Cambridge University Press.

Selfridge, O. 1959. Pandemonium: A paradigm for learning. Proceedings of the 10th Symposium of the National Physical Laboratory, vol. 1 D. V. Blake and A. M. Uttley, editors. pages 511–529. Publication of 1958 Symposium

Shepard, R. 1962. "The analysis of proximities: multidimensional scaling with an unknown distance function." *Psychometrika* 27:219 246.

Sims, K. 1994. "Evolving Virtual Creatures." *Computer Graphics.* Proceedings of Siggraph 1994, 15–22.

Tenenbaum, J. B, T. Griffiths, and C. Kemp. 2006. "Theory based Baysian models of inductive learning and reasoning." *Trends in Cognitive Science* 10:309–318.

Tinbergen, N. 1951. *The Study of Instinct.* Oxford: Clarendon Press.

Tononi, G. 2008. "Consciousness as integrated information: a provisional manifesto." *Biology Bulletin,* 216–242.

Von Neumann, J and O. Morgenstern 1944 *Theory of Games and Economic Behavior* Princeton University Press

Watts, D. J. 1999. *Small Worlds: The Dynamics of Networks between Order and Randomness.* Princeton: Princeton University Press.

Watts, D., and S. Strogatz. 1998. "Collective dynamics of small-world networks." *Nature* 393: 440–442.

Wiener, N. 1948. *Cybernetics: Or Control and Command in Animals and Machines.* Malden, MA: Wiley & Sons.

Wilson, E. O. 1971. *The Insect Societies.* Cambridge, MA: Harvard University Press.

Wilson, E. O. 1998. *Consilience: The Unity of Knowledge.* New York: Knopf.

Wolfram, S. 2002. *A New Kind of Science.* Champaign, IL: Wolfram Media.

Xie, X-H, R. Hahnloser, and H. S. Seung. 2001. "Learning winner-take-all competitions between groups of neurons in lateral inhibiting networks." *Advanced Neural Information Processing* 13:350–356.

Young, H. P. 1988. "Condorcet's Theory of Voting." *American Political Science Review* 82:1231–1244.

Young, H. P. 1995. "Optimal Voting Rules." *Journal of Economic Perspectives* 9:51–64.

Young, H. P. 1998. *Individual Strategy and Social Structure.* Princeton: Princeton University Press.

Zipf, G. K. (1932) *Selected Studies of the Principle of Relative Frequency in Language.* Cambridge, MA: Harvard University Press.

Appendix: Phase Plots

A phase plot is a planar cut through the n-dimensional space of potential winners for a set of weights on vertices of the graph G_n. The cut shows the Condorcet winners when two of the vertices have variable weights, and all the remaining vertices have fixed weights. The java program for generating these plots was written by T. J. Purtell in 2004 when he attended my anigrafs class.

In T. J. Purtell's construction, the initial topology for G_n is a circular ring of vertices labeled clockwise from the node at 3:00. For display purposes, colored labels are assigned to the vertices. These are chosen from the color circle beginning at red and progressing through yellow, green, etc., to magenta: {R, O, Y, YG, G, C, B, V, P, M}. The hue list is a continuous function, and the color increments divide the circle into n equal steps. In the interface display, the lower panel includes setting for the knowledge depth (Kd), which is here set at 2, while the maximum value for weights is set at 6 (see cursor above scale). If Kd = 2, then the calculation of weights for any node uses the node weight plus that of its neighbor(s) in the graph.

Two of the weights are given variables x and y, as indicated. Their origin is at the top left of the phase plot. As x and y are increased, the Condorcet outcomes are displayed as colors.

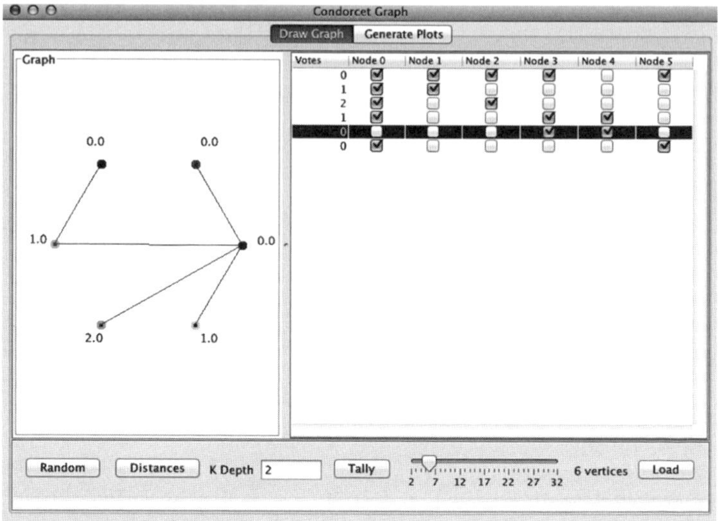

Figure A.1

A region of uniform color indicates the winning vertex for that set of x, y weights, plus the initial weights on the other n–2 vertices. If there are top-cycles, these appear as textured regions, with the colors showing the two dominant members of the top-cycle.

A phase plot for the tree graph example is given in the right panel. This is the form of the output from Purtell's program. There is one textured parallelogram showing a region of top-cycles. As mentioned in the text, the boundaries between regions are straight lines where a change in the weights on alternatives created a reversal in an inequality in pairs of alternatives being compared. In the phase plot, we thus know that two of the members of the top-cycle, node 1 and node 5, are red and blue, respectively. The missing member of the top-cycle can be found as follows:

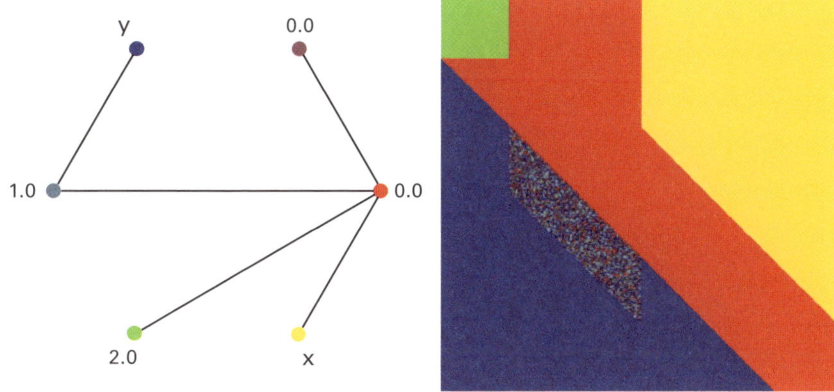

Figure A.2

For clarity, relabel the nodes 1 to 6 as A through F. For each of the two alternatives already recovered (red node 1 = A, blue node 5 = E) calculate the weight relations for all pairs B through F, and likewise for alternative E. For both A and E, there will be one pair that the alternative does not dominate. For weights on x and y, note that a rough estimate of the middle of the textured region will be $x = 2$, $y = 3.5$. Use these values to calculate the inequalities for Kd = 2. These values are:

Pairs for A: **AB**=3 versus 2, **AC**=3 versus 2, **AD**=4 versus 4.5, **AE**=4 versus 3.5, **AF**=5 versus 0

Pairs for D: **DA**=4.5 versus 4, **DB**=4.5 versus 2, **DC**=4.5 versus 2, **DE**=1 versus 3.5, **DF**=4.5 versus 0

The winner of a pair is highlighted in bold. Note that D beats A, and, in the second comparison, E beats D. Hence the top cycle is E > D > A > E.

We can now easily calculate equations for the boundaries of the top-cycle:

A > E = 2 + x > y; D > A = y + 1 > 2 + x. **The left vertical boundary is at C = A;**

hence 2 = x + 1, or x = 1. The right boundary is an extension of A = B, or 3 = x. Continuing this procedure for different slices through the phase space, we can reconstruct the boundaries of the top-cycle polyhedron, and thus calculate deterministically the probability of the top-cycle.

Glossary

adjacency matrix: The rows of $A = [e_{ij}]$ correspond to the n vertices v_i of the unlabeled graph G_n, and the n columns are all the possible neighbors v_j. If there is an edge from v_i to v_j, then $e_{ij} = 1$; otherwise $e_{ij} = 0$.

agent: A decision-making entity having specific goals and preferences.

algebra: A formal system for relating and operating on elements of abstract structures, such as number systems or groups.

anigraf: A social network of mental organisms or agents. Refer to Preliminaries, section 4.

atoms: Connected subgraphs associated with the binary decompositions of the edge sets. Refer to section 9.3.1

automorphism (of G):   A one-to-one mapping f of the vertex set $V(G)$ onto itself with the property that f (v) and f (w) are adjacent if, and only if, v and w are adjacent.

bipartite: $K(m,n)$ is a graph Gmn where the vertices can be partitioned into two unconnected subsets such that every edge of G joins vertices in one set with vertices in the other.

Borda* count: Preferences are weighted inversely to their rank to determine the voting power of an individual's choices. Specifically, if there are three choices, the Borda* weight vector will be {2, 1, 0}. This vector will be applied respectively to the first, second, and third choices of each voter, and then these weights will be summed for all voters to give the Borda* count. Note that, unlike the classic Borda count, choices at the same level in the preference order are all given equal rank weight. Refer to Anigraf4.

broker: A high-level mental organism of an anigraf that controls sequences of actions by agents.

causal network: A directed acylic graph that represents the history of the anigraf's evolution.

centroid: That vertex with the highest degree.

chain: C_n is a chain if all vertices have degree two, excepting the two (ends), which have degree one.

chaos: An irregular and unpredictable sequence of states of a system such that past behaviors are not repeated. Typical examples include nonlinear mechanical oscillators, fluids, and some chemical reactions. However, chaos can also occur in social decision-making.

circumference: The length of the longest graph cycle.

clique: A subgroup of an anigraf population with similar goals.

coelenterate: Invertebrate animals, such as sea anemones, jellyfish, and hydroids.

coevolution: The evolution of different anigraf species whose development depends critically on the interactions among the species.

cognitive modes: Frameworks for knowledge, thought, and especially reasoning by analogy; more generally, constraints on theories of natural modes. See also *Ideonomy.*

complement: The complement of G_n has all the remaining edges of the complete graph K_n, and none of the edges of G_n.

conditional probability: $p(A/B)$: the probability of event A, given that event B has occurred.

Condorcet winner: The alternative (goal, choice) that beats all other alternatives in a pairwise contest.

coordinate frame: A distinguished point (the origin of the frame) with axes representing the different variables, with categories such as numbers assigned to positions along the axes.

covered graph: If at least one vertex is adjacent to all the remaining vertices, the graph will be covered. Covered graphs will always have a Condorcet winner.

cutpoint: A vertex is a cutpoint of a connected graph if the removal of this vertex creates a disconnected graph. If the removal of an edge creates a disconnected graph, the edge is called a *bridge.*

cybernetics: From the Greek word for "steersman." Refers to the field of feedback control and communication theory, whether in the machine or in the animal (N. Weiner 1948).

cycle, graph: An alternating sequence of $n \geq 3$ distinct vertices and distinct edges linking these vertices, with the last member of the vertex sequence also being the first, thus creating a closed path.

cycle, social: Given at least three alternatives, a sequence of winners in a pairwise (Condorcet) contest, such that the last member of the sequence beats the first. See also *top-cycle*.

daemon: A mental organism.

dance: Typically, a sequence of movements, such as gaits, executed in coordination by two physically distinct entities. However, it can be generalized to include the interplay between two mental organisms.

degree: The number of edges incident with a vertex.

diameter: The longest length of the shortest path between any two vertices in a graph.

digraph: D_n is a graph G with directed edges between vertices. Also called a *directed graph*.

dimensional analysis: A method that utilizes the dimensions of variables used to describe natural phenomena in order to recover the form of the underlying model. Assumes that the phenomenon can be described by a dimensionally correct equation among the particular variables. Dimensionless products of variables and linear algebra play a key role.

directed graph: See *digraph*.

dispersion game: A game with an outcome that maximally disperses agents over a set of possible actions. Useful for optimizing the roles of players on team, where actions and talents of the players differ.

domain map: See *knowledge structure*.

dynamical system: A smooth transformation of variables that characterizes the evolution of the state of the system. For example, $x_{n+1} = f(x_n)$.

eccentricity (of a vertex):   The longest length of the shortest path from the chosen vertex to any other.

euglena: A one-celled organism up to 0.01 inch long with a "tail" (flagellum) that whips to provide locomotion. Often found on the surface of ponds, where chlorophyll used for photosynthesis gives the organism a greenish color.

fixed point: The terminal point of an iterated function or map. For anigrafs, also an equilibrium or stable point.

flagellum: A long, thin fiber that moves like a whip to propel an organism.

frame: See *coordinate frame.*

free agent: Mental organisms whose preferences are initially unassigned, but are capable of acquiring a new goal state and learning its relation to other goals already in place.

gait: A style of animal locomotion. For legged creatures, examples are a walk or gallop for a horse, or a tripod gait for a cockroach.

game theory: A mathematical theory of bargaining first outlined in detail by von Neumann and Morgenstern, 1944.

General Possibility theorem (Arrow, 1951):   Impossibility of universal choice rules for total preference orders, given five very plausible assumptions: nondictatorship, collective rationality, Pareto (unanimity), independence of irrelevant alternatives, conflict resolution.

girth: The length of the shortest graph cycle.

grammar, graph: A set of rules or productions (and possibly prototypes or motifs) that create or transform one graph into another.

graph (undirected): G_n is a finite non-empty set of n vertices (points, nodes) with up to $_nC_2$ edges (lines, arcs) between the n vertices.

graph (isomorphic):   G and H are isomorphic if they have a one-to-one correspondence between their vertex sets that preserves adjacencies.

graph (labeled):   See *labeled graph.*

graph (line):   See *line graph.*

graph (ring):   See *ring.*

groups: The permutation group of G are the automorphisms of G.

ideonomy: A term coined by Patrick Gunkel to describe the study of laws underlying conceptual spaces, their manipulations, and their extensions. Refer to ideonomy.mit.edu.

induced subgraph: For $k < n$, if all k vertices and edges of g_k are in G_n, then g_k is a subgraph of G_n. The induced subgraph $< g_k >$ is one where two points in g_k are adjacent if and only if they are adjacent in G_n.

information (bits):   Shannon information is $[(1/p)\text{Log}_2(p)]$, where p = event probability. For anigrafs, bits can also be used to measure the informativeness of an event, namely the extent to which an anigraf model is revised or updated, and hence a measure of meaningfulness.

IRM: Innate releasing mechanism.

isomorphic: Two graphs are isomorphic if their vertex and edge sets are identical.

knowledge depth (Kd): A number two fewer than the maximum number of levels in the partial ordering of a voter's preferences. If tallies are conducted using the ideal point (first choice) and only its neighbors in G_n, and all other alternatives third-ranked and higher are weighted zero, then Kd = 2.

knowledge structure: A model (graph) of the similarity relations among alternatives. If the knowledge structure is *shared*, then all possible individual preference rankings are consistent with the model.

labeled graph: A graph with its vertices distinguished from one another by names.

landmark: The intersection of at least two causal paths generated by two independent parameters.

Laplacian matrix: For the undirected graph G_n with n vertices vi, each assigned to one row and column, the (symmetric) entries are defined by d_i if $i = j$; -1 if $i = ! = j$ and there is an edge (i, j); otherwise 0, for $i, j = 1 \ldots n$, where d_i is the degree of v_i.

likelihood: In context, C, the probability of a particular observation, o_i, given the presence of an event e_i; expressed as $p(o_i/e_i,C)$.

line graph: The vertices of the line graph L(G) of G consists of the edges of G, with two vertices in L(G) adjacent whenever the corresponding edges of G are adjacent. Note that every cutpoint of L(G) will be a bridge in G.

multidimensional scaling (MDS): A technique for mapping similarities among items (Shepard 1962).

mental organisms: Agents or daemons in a mind or cognitive system, here represented as nodes in a graph that shows the communication links and similarities of goals of the constituents of the cognitive system.

metagraf: A relationship among a set of anigraf models. These higher order representations include transformations and dynamics, cast as graphs.

model: Given a representational system, a model fills out that representation with assignments of variables or particular relations that make explicit how the representation can be used.

multiscale graph: A graph with a fractal distribution of the shortest mean path lengths from one vertex to all others. See also *scale-free graph* and *small world graph*. The latter have been shown to be multiscale (Kasturirangan 1999.)

natural mode: A correlated set of quasi-independent rules or regularities describing a natural phenomenon. The presence of one regularity is highly predictive of the other regularities.

network: Here used almost synonymously with a connected graph, with the typical network being a very large (random) graph.

pandemonium: The name Oliver Selfridge gave in 1959 to a fanciful pattern recognition system where: feature daemons shouted evidence for types of image features, which were then heard by cognitive daemons who were each listening for evidence supporting particular patterns; these daemons in turn called out to a decision daemon who made the final choice.

path: An alternating sequence of distinct vertices and distinct edges in which each edge is incident with the two vertices immediately preceding and following it in the sequence.

phase plot: For anigrafs, a plot showing regions of Condorcet winners and regions of top-cycles, if present. Refer to Appendix: Phase Plots.

planar graph: When drawn in a plane, no two edges meet (or cross) except at a vertex in which both are incident.

preference order: A partial ordering of alternatives that makes explicit an individual's first, second, third, etc., choices among those alternatives.

priors: The probability of events at the beginning of a causal chain of inference. For beliefs, the priors summarize information or knowledge before the beliefs are updated.

Prisoner's dilemma: A game in which players (partners in crime) may defect (confess) or cooperate with each other (don't confess) on each trial. There are payoffs to each player depending on their choices. For two players, the lowest payoff is to the one who cooperates when the other defects, who then gets the highest payoff. The next lowest payoff is when both defect, and the second highest payoff is when both cooperate.

proxy: An agent or mental organism that resides in one anigraf, but votes the wishes of another.

radius: The minimum eccentricity of the vertices.

Ramsey numbers r(G,H):   Given two graphs, G and H, r is the smallest number such that if the edges of the compete graph K_r are colored red and green, then there is either a set of red edges forming a subgraph isomorphic to G or a set of green edges forming a subgraph isomorphic to H (Read and Wilson 1998).

regular graph: All vertices have the same degree.

representation: A formal system for making explicit certain entities or types of information, together with a specification of how the system does this (Marr 1982). Note that the representational form plays an important role in how variables are manipulated. A simple example is the Arabic, Roman, and binary systems for representing numbers.

ring: A closed chain; a regular graph with all vertices of degree two.

rooted tree: A tree graph with one vertex distinguished as the origin.

Sarkovskii's lemma (theorem):   A cycle of period three in a finite attractor for quadratic maps implies periodic points with all periods.

scale-free graph: A graph with the distribution of k vertex degrees that obeys a power law of $p(\mathrm{k}) = \mathrm{ck}^{-\lambda}$.

sensory order: Hayek's (1952) view that perception should be regarded as classification of sensory data, thus establishing an order to these data. Similarly, the mental order of events in an environment would be the internal principles used to place an ordering on relations attributed to these events. Note the underlying natural modes assumption, which relates physical orders to sensory orders.

similarity: The percent of vertices in two graphs having identical adjacencies. Relevant to the similarity of Condorcet outcomes and information measures.

small world graph: "Small world" refers to the observation that there are short paths of connectivity between any two individuals, thought to be about six for the optimal iteration of mutual acquaintances (hence dubbed "six degrees of separation"). A small world graph is highly clustered and yet has rather a short median distance for the means of the shortest path lengths connecting each vertex to all other vertices.

social network: A graph showing the connectivity relations or interactions among individuals, groups, or nations, etc.

social order: S_n is the single partial ordering of alternatives that is the outcome of a voting procedure that aggregates a finite number of weighted partial orderings of these same alternatives. More simply, the ranking of all alternatives based on the results of a vote, with the winner ranked first, runner-up second, etc.

spectrum (of a graph):   The Eigen-values for the characteristic polynomial describing a graph. Refer to Read and Wilson 1998.

storyline: A directed causal graph depicting a sequence of actions and events carried out by at least two players. Simplest story is: boy meets girl; a negative force comes into play and boy loses girl; the negative force is overcome, and boy is reunited with girl.

subgraph (induced):   For $k < n$, if all k vertices and edges of g_k are in G_n, then g_k is a subgraph of G_n. The induced subgraph $< g_k >$ is one where two points in g_k are adjacent if and only if they are adjacent in G_n.

tit-for-tat: A strategy for cooperation or defection in iterated games such as the Prisoner's dilemma: namely, cooperate on the first move, and then do whatever the other player did on the preceding move.

trajectory mapping (TM):   A technique for mapping paths of similarity relations among items (Richards and Koenderink 1990).

top-cycle: A set of k alternatives a_i, with $k > 2$, such that a_1 beats a_2 beats $a_3 \ldots a_{k-1}$ beats a_k and a_k beats a_1; also, every alternative not in the top-cycle is beaten by at least one alternative in the top-cycle.

tree: A graph with no cycles.

vector: A set of objects, typically a set of scalars, with rules for addition (commutative, associative) and scalar multiplication. In R^3 a vector is usually defined as a directed line between pairs of triples (x, y, z).

vehicle: Braitenberg's (1984) creation of a robot-like machine with a very simple internal structure, equipped with sensors and motors for interactions with its environment. Their impact is such that one is tempted to use psychological language to describe a vehicle's behavior.

wheel: (W_n) the ring graph R_{n-1} with an extra point that covers all $n-1$ vertices in the ring.

Index